THE IQ GAME

THE IQ GAME

A Methodological Inquiry into the Heredity-Environment Controversy

Howard F. Taylor

Rutgers University Press
New Brunswick, New Jersey

Permission to reprint the following material is gratefully acknowledged.

Figures 1 and 2 and Tables 2, 3, and 4 from D. C. Rao, N. E. Morton, and S. Yee, "Resolution of Cultural and Biological Inheritance by Path Analysis," American Journal of Human Genetics 28 (1976):229–233. Reprinted by permission of The University of Chicago Press. © 1976 by the American Society of Human Genetics. All rights reserved.

Figure 1 from R. C. Lewontin, "The Analysis of Variance and the Analysis of Causes," American Journal of Human Genetics 26 (1974):405. Reprinted by permission of The University of Chicago Press. © 1976 by the American Society of Human Genetics. All rights reserved.

Table 2 from A. R. Jensen, "How Much Can We Boost IQ and Scholastic Achievement?" Harvard Educational Review 39 (Winter, 1969). Reprinted by permission. Copyright © 1969 by President and Fellows of Harvard College.

Table 3 from D. C. Rao and N. E. Morton, "IQ as a Paradigm in Genetic Epidemiology," in N. E. Morton and C. S. Chung, eds., Genetic Epidemiology (New York: Academic Press, 1978), p. 156. Reprinted by permission.

Figure 1 from L. Erlenmeyer-Kimling, "Gene–Environment Interactions and the Variability of Behavior," in L. Ehrman et al., eds., Genes, Environment, and Behavior: Implications for Educational Policy (New York: Academic Press, 1972), p. 190. Reprinted by permission.

Figure 1 from L. Erlenmeyer-Kimling and L. F. Jarvik, "Genetics and Intelligence: A Review," Science 142 (13 December 1963):1477–1479. Reprinted by permission. Copyright 1963 by the American Association for the Advancement of Science.

Tables A-6, A-7, A-8, A-9, and A-10, and Figure A-3 from C. Jencks et al., Inequality: A Reassessment of the Effect of Family and Schooling in America (New York: Basic Books, 1972), pp. 286–287, 289, 291, 293, 294, and 312. Reprinted by permission of Basic Books and Penguin Books Ltd.

Library of Congress Cataloging in Publication Data

Taylor, Howard Francis, 1939–
   The IQ game.

   Bibliography: p.
   Includes index.
   1. Intellect.   2. Nature and nurture.   I. Title.
BF431.T254   155.2'34   80–13997
ISBN 0–8135–0902–5

To Carla

Contents

	Preface	ix
Chapter 1	**Galton's Legacy**	1
	A Historical Note	7
Chapter 2	**Estimating IQ Heritability**	10
	The Coefficient of Heritability	10
	Estimating IQ Heritability	16
	The Separated Identical Twin Correlation	18
	Kinship Correlation	25
	The Cyril Burt Scandal	35
	Jensen's Kinship Equation via Jencks	41
	A Consistency among the Inconsistencies	60
	A Note on Race Differences and Between-Group Heritability	64
	Gene-Environment Covariance and Interaction	68
	Summary	72
Chapter 3	**The Myth of the Separated Identical Twins**	75
	The Samples and Sample Bias	78
	A Close Look at the Data	84
	Late Separation	85
	Reunion Prior to Testing	87
	Relatedness of Adoptive Families	90
	Similarity in Social Environment	92
	A Note on Measurement Error	101
	Jencks et Al.'s Path Estimates Using Separated Identical Twins	103
	Summary	109

Chapter 4   **Studying Kinships Two at a Time**   112
             The Requirements of Heritability
                Estimation   112
             A Look at Path Analysis   119
             A General Collateral Kinship Model   124
                Measurement Error   130
                Separated Identical Twins   134
                Adopted Children: Unrelated Pairs
                   Raised Together   135
             Kinship Correlations and Bogus
                Heritability   137
                Pairwise Comparison of Kinship
                   Categories   148
             Gene-Environment Interaction   154
                Nonadditive Path Models: An
                   Illustration   162
             Summary   166

Chapter 5   **Studying Kinships Many at a Time**   171
             Solving Three Equations: An Example
                from Jensen   171
             The Honolulu Research   175
                The Honolulu Equations and a
                   Modification   187
                An Illustrative Analysis   192
                Some Conclusions   198
                The Honolulu Data Sets   200
             A Note on the Birmingham Models   202

Chapter 6   **Summary and Conclusions**   205

Appendix A   **The Cognitive Style Challenge to IQ, by**
                D. R. Vasgird   218

Appendix B   **Coded Data on Separated Identical Twins**   226

             Notes   229

             References   244

             Index   262

Preface

The "nature-nurture" controversy has held a prominent place in scientific controversy for over one hundred years. The essential issue can be stated now in virtually the same terms as it was stated then: Are human differences in character and behavior influenced more by biological heredity (nature), or by social and educational environment (nurture), or perhaps by some combination of both? The present-day version of this controversy centers on intelligence testing: Are differences among people on IQ tests determined more by their genetic differences or by their environmental differences? This book is about that question. To be sure, I touch on related matters such as whether or not any race differences in IQ could have a genetic basis, and whether or not the nature-nurture question with regard to IQ is even scientifically answerable.

Much of this book will be understandable to the nontechnical reader. Chapters 1, 2, 3, and 6 require little knowledge of behavioral science research methodology or statistics. A knowledge of multivariate statistical procedures will be helpful, though not absolutely necessary, for the material in Chapters 4 and 5. Those already familiar with multiple regression and path analysis will find Chapters 4 and 5 easy reading. Chapter 6 provides a nontechnical and brief summary of everything covered in the book.

Numerous individuals have in a variety of ways contributed to the completion of this work. I particularly wish to thank Marsha Bonner, Jamesetta Reed, Reginald Moore, Kassandra Hannon, and Michael Jackson for their assistance in library research during the early and intermediate phases of this project. I thank Robert Porter of Princeton's department of economics for his programming and computational assistance in connection with the weighted least squares analyses in

Chapter 5. I thank M. Stephen Sasaki of Princeton's department of sociology for his ever-present advice and assistance. Work at various
stages was partially supported by a grant from the Russell Sage Foundation, and a grant to Princeton's Afro-American Studies Program
from the Rockefeller Foundation. The sociology department and the
Afro-American Studies Program of Princeton University deserve special thanks for further support at various stages. Writing of the final
manuscript was supported by a grant from the Maurice Falk Medical
Fund. Philip Hallen, president of the Maurice Falk Medical Fund, has
supported the completion of this work in many valuable ways. Marlie
Wasserman of Rutgers University Press nurtured the entire project
from start to finish.

I thank the following persons for alerting me to sources and for wel-
come advice and help of various sorts: Jack Stavert, Thelma Spencer,
Theodore L. Cross, Carlton Hornung, Christine LeFlore, Angela March,
Mary-Helen Thompson, June Murray-Gill, Milton Rokeach, and
Charles V. Willie. Members of a special Carnegie Panel deserve special
thanks for whetting much of my initial interest in doing this book:
Andrew Billingsley, James Comer, Ron Edmonds, William Hall, Robert
Hill, Nan McGehee, Lawrence Reddick, and Stephen Wright.

The following have graciously provided me with portions of their
work prior to publication: Arthur R. Jensen, William Shockley, N. J.
Block, Gerald Dworkin, William Hall, Peggy Sanday, Newton E. Morton, Paul Taubman, Rick Heber, Sandra Scarr, Leon Kamin, Christopher Jencks, Michael Schwartz, and Arthur Goldberger. I thank Professors Scarr, Kamin, and Goldberger and Steven G. Vandenberg, L.
Erlenmeyer-Kimling, Richard Herrnstein, Robert Gordon, and D. W.
Fulker for responding to my inquiries about their work. I am indebted
to Professors Jencks and Kamin and to Mark Abrahamson, William
McCord, Marshall Segall, Byron Egeland, and John Loehlin for reading
and commenting on various portions and drafts of the manuscript. I
express special thanks to Allan Mazur, Robert Hamblin, Arthur S.
Goldberger, and Michael Schwartz for particularly detailed and helpful comments. Any errors and misjudgments on the pages that follow
are mine, not theirs. A large intellectual debt rests with my teacher,
colleague, and friend, H. M. Blalock, Jr., for it was he who initially
alerted me to being constantly aware of the role of assumptions in the
analysis of data, an awareness that I hope is appropriately evident in
the pages that follow.

Hattie Black and Comfort Sparks of Princeton University deserve
special thanks for typing and proofreading, respectively, virtually the

entirety of all the various drafts and versions of this work. Marie Ann Pavlak typed portions of the final draft, and Geraldine Reed provided clerical assistance. A special debt of love and gratitude goes to my wife, Patricia E. Taylor, Esq., and my daughter Carla Y. Taylor, who once again bore with the vicissitudes and late hours of a preoccupied husband-father. A continuing inspirational debt rests with our friends Elihu and M. Sue Gerson—and I once promised them I would say this in precisely these terms—without whom this book would not have been possible.

Far better an appropriate answer to the right question, which is often vague, than an exact answer to the wrong question, which can always be made precise.

J. W. Tukey

Of all vulgar modes of escaping from the consideration of the effect of social and moral influences on the human mind, the most vulgar is that of attributing the diversities of conduct and character to inherent natural differences.

J. S. Mill

What heredity can do environment can also do.

Newman, Freeman, and Holzinger

Epaminondas, oh Epaminondas!
You ain't got the sense you was born with.

from an Afro-American folk tale

Chapter 1

Galton's Legacy

Few treatises appearing in otherwise prosaic scholarly journals have caused so much controversy over the years both inside and outside academia as the now-famous article by Arthur R. Jensen, in the Winter 1969 issue of the *Harvard Educational Review*, "How Much Can We Boost IQ and Scholastic Achievement?" Jensen presented the argument, heard periodically through history from pioneer statistician Sir Francis Galton to the present, that intelligence, to the extent that it can be measured, is determined more by genetic makeup or "nature" than by social environment or education ("nurture"). Echoing a conclusion he drew in a seminal article two years earlier (Jensen, 1967), Jensen concluded that the *genetic heritability* of human IQ for populations of white individuals is "substantial," approximately .80, meaning that approximately 80 percent of all differences in human IQ is attributable to some kind of hypothetical genetic mechanism, thus allocating about 20 percent to such factors as social environment, education, gene-environment interactions, and everything else. Thus according to Jensen IQ is genetically determined to roughly the same degree as certain physical attributes, such as height—a characteristic also thought to have a genetic heritability of 80 percent. Jensen feels that the genes-to-IQ connection is so strong that "we can, in fact, estimate a person's *genetic standing* on intelligence from his score on an IQ test" (Jensen, 1970a, p. 131, italics added).

Although Jensen's conclusions about the heritability of IQ within white populations stimulated furor enough, what he hypothesized about racial differences in intelligence catapulted the nature-nurture issue into the limelight of contemporary history. Jensen noted that blacks and whites differ on the average by roughly 15 points on standard IQ tests such as the Stanford-Binet, Wechsler, Lorge-Thorndike,

1

and Raven's Progressive Matrices. He then reasoned that if IQ is largely determined genetically, it follows that it is a "not unreasonable hypothesis that genetic factors are strongly implicated in the average Negro-white intelligence difference" (Jensen, 1969, p. 82). He observed that races are "breeding populations" in the sense that child-producing matings within a racial group have a higher probability of occurring than child-producing matings between racial groups. He surmised that because certain apparent and presumably genetically based physiological differences appear between races (skin pigmentation; blood-group differences; many other physical differences), there is "no reason to suppose that the brain should be exempt" from such differences, and thus there is *"little question"* that "racial differences in genetically conditioned behavioral characteristics, such as mental abilities, should exist, just as physical differences" (Jensen, 1969, p. 80, italics added).

Virtually all of Jensen's writing over the years since his 1969 article has continued to favor substantial genetic heritability for IQ. For example, in 1970 he concluded that the evidence "is very clear" that there is "a large genetic component in individual differences in IQ" (1970, p. 130); in 1971 that differences in "mental abilities" have "a substantial genetic component" (1971, pp. 40–41). In 1972 he asserted confidently that it is a "scientifically established fact" that "hereditary or genetic factors account for more of the IQ differences among persons than do cultural and environmental factors" (1972a, p. 14). In 1973 he observed, "I have found nothing to cause me to alter my original thesis in the 1969 *HER* article in any major respect" (1973, pp. 22–23); in 1975 that IQ heritability is "from about .60 to .90" (1975, p. 172); and in 1978 from ".65 to .80" (1978b, p. 14).

Since 1969, Jensen has continued to develop and refine his conclusions regarding race differences in IQ, stating in 1972 that while genetically based differences *within* either white or black groups do not "prove" that any average difference *between* groups is attributable to genetic factors, it is nevertheless true that "the higher the heritability of a trait within each of two groups, the greater is the likelihood that a mean difference between groups has a genetic component" (1972, p. 29). Indeed, in 1973 he felt confident enough to venture the quantitative estimate that "something between one-half and three-fourths" of the black-white difference in IQ "is attributable to genetic factors" (1973, p. 363). More recently, and only slightly more cautiously, he states there is a "considerable *plausibility* of there being some nontrivial genetic component in the IQ difference between racial groups" (1978b, pp. 22–23).

This book is devoted to a detailed analysis of Jensen's methods and data. To be sure, it also focuses in some detail on the related analyses of others. I have kept in mind two central questions: Do the methods and data of Jensen and the others permit the inference that human IQ is "substantially" genetically determined? Indeed, do these methods and data permit any inferences at all about the magnitude of IQ heritability?

The question is not merely academic. Jensen's conclusions have been construed—correctly or incorrectly—as having implications for educational and social policy, particularly as such implications bear upon the treatment of minorities.[1] Whether or not such research (and the conclusions drawn therefrom) *should* bear upon social policy appears at least in part irrelevant, since it seems clear that hereditarian research has already had at least some measurable impact on important aspects of educational policy.[2]

Jensen's work has provided a considerable stimulus for William Shockley, physicist, winner of the Nobel Prize, coinventor of the transistor, and Alexander M. Poniatoff Professor of Engineering Sciences (Emeritus) at Stanford University, who since 1964 in numerous publications, memoranda, letters, TV appearances, speeches, and proposals before the National Academy of Sciences, has repeatedly asserted not only that "the principal cause of our American Negro problems is racially genetic" (Shockley, 1970, p. 4) but in addition—with clear genocidal overtones—that low-IQ blacks might be "voluntarily" sterilized to improve the "quality" of the overall population (1973, p. 4). This "voluntary sterilization bonus plan" proposes that participants receive "a thousand dollars for each point below 100 IQ, 30 thousand dollars put in trust for a 70 IQ moron, potentially capable of producing 20 children" (1972, p. 17). Shockley has repeatedly made the utterly unsupportable claim that "the average IQ of Negro populations increases by about one IQ point for each 1% of added Caucasian genes" (1969, p. 2; 1972, p. 16). Shockley attributes his own "scientific" concern with the sterilization of Afro-Americans to an incident that occurred in 1963—one year before his first National Academy of Sciences proposal—in which a black "low-IQ" teenager threw acid in the face of a delicatessen proprietor. He concluded from this that "such incompetent people were anti-evolutionary" (1973, p. 2). His later pronouncements are no more encouraging to Afro-Americans, for he asserts that those who are "born in slum environments [have] statistically poor heredity from unfair shakes from the badly-loaded genetic dice cups of their parents" (1978, p. H-3).

Jensen, who writes primarily for an audience consisting of professional social scientists and students, has found backers in the popular press. One is Professor Richard Herrnstein, two of whose publications are of interest—an article appearing in the September 1971 issue of the *Atlantic* and a book published in 1973 that expands on the article. In both works, following Jensen closely, Herrnstein develops four main points:

1. Currently used IQ tests, such as the Stanford-Binet test and the Wechsler Adult Intelligence Scales (WAIS), do indeed constitute valid and reliable measures of "intelligence."
2. Intelligence is approximately 80 percent genetically heritable. In this connection, he makes the rather bold claim that "one can search the scientific literature from one end to the other, as I have now done, and find no significant empirical challenge to the sizable genetic contribution to scores on intelligence tests" (1973, p. 9). (Herrnstein's search must have been considerably more modest than his claim suggests, for several well known studies [reviewed in detail later] published prior to Herrnstein's book, notably those of Scarr-Salapatek, 1971a, and Jencks et al., 1972, found somewhat less than a "sizable genetic contribution" to how people respond to "intelligence" tests. [Jencks et al. found IQ heritability to be about .45.] Furthermore, there is evidence that Herrnstein simply misread and misinterpreted some of this literature which he claims to have searched so completely.)[3]
3. Since IQ scores depend heavily upon genes, and since "success" in society (including earnings and social prestige) depends largely upon IQ, it follows that success in society is determined largely by genes. This last point, which has come to be known as "Herrnstein's syllogism," is central to his final point.
4. If individuals are made to become more equal in their social environments, as—so he argues—is one intent of our present educational system, then even less of the individual differences (variance) in IQ will be explained by environment, and hence even more than 80 percent of the individual differences in IQ will be explained by genes.

Our society is thus, according to Herrnstein, moving inexorably toward an IQ-based, genetically determined, castelike "meritocracy," with the genetically disadvantaged at the bottom and the genetically more fortunate at the top. Unlike Jensen, Herrnstein draws no conclusions about between-race differences in intelligence; he nevertheless concludes that social *classes* differ in genetically caused intelligence—the higher classes having the more "intelligent" genes. In fact, Jensen himself has subscribed to the Herrnstein meritocracy principle, stating that the "less able gravitate downwards in the SES [socioeconomic status] hierarchy. In so doing, they of course take their genes for intelligence with them" (1973, p. 152).

I also examine the work of other investigators of IQ heritability. One is H. J. Eysenck, a psychologist whose book, published in England as *Race, Intelligence, and Education,* was subsequently published in the United States under the title *The IQ Argument* (1971). The American edition was lauded by Jensen as an "admirably lucid" account of race differences in intelligence (1972, p. 59). Eysenck concludes that IQ is approximately 80 percent heritable for whites (1971, p. 57 and passim) and probably for blacks as well (p. 67 and passim). He has continued to argue in favor of 80 percent heritability even in his more recent statements (e.g., in the *Times* of London, November 12, 1976). He argues generally that black-white differences in intelligence indeed exist and are "inevitably" largely genetically determined (1971, pp. 79, 85, 95–96, 108, and passim). Quoting an earlier controversial but less well known work by Shuey (1966), Eysenck maintains that the facts "all taken together inevitably point to the presence of native differences between Negroes and whites as determined by intelligence tests" (1971, p. 108). In fact, Eysenck surpasses Jensen by a considerable margin in implicating other ethnic groups in the United States and England—"Italians, Spaniards, and Portuguese, as well as Greeks," as "less able, less intelligent" and as "very poor samples" of the populations of their countries of origin (p. 43). Just before his death, the late Cyril Burt, around whom a major scandal now rages in the scientific community (summarized later), said of Eysenck's book that it "is as sound as it is impartial" (quoted on the book jacket of Eysenck, 1971). Jensen—who was a postdoctoral student of Eysenck's—feels that his mentor's book is "in the best tradition of popular science writing" (1972, p. 59), a claim difficult to understand.[4] I examine significant portions of this "best" in the following chapters, and in fact take direct issue with one of Eysenck's proclamations, that his critics "seem to fight shy of the technical literature" (Eysenck in the *Times* of London, November 8, 1976). We can assure Professor Eysenck that we will be taking a look at the technical literature.

While Jensen, Herrnstein, and Eysenck (and certainly Shockley) all conclude that the genetic heritability of IQ is substantial, other investigators have taken more moderate stances—although sometimes only slightly more moderate. In 1967 Bruce Eckland, a sociologist, argued that IQ is determined more by genes (more than 50 percent) than by social environment (p. 178); in 1971, much like Herrnstein, that intelligence differences between social classes have "partly a hereditary basis" (1971, p. 65); and later, citing Jensen, that genes play "a very powerful role" in determining IQ (1973, p. 86). I also consider in detail the work of another sociologist, Christopher Jencks, and his associates at Harvard (Jencks et al., 1972), who differed with Jensen and concluded, after an elegant and mathematically detailed analysis, that the heritability of IQ is approximately 45 percent, with 35 percent of the variance going to environment and the remaining 20 percent to what is called the covariance (correlation) between genes and environment. The work of Jencks and associates, which employs path analysis, a technique of considerable popularity in social science research, has been called "the best in the literature" (Stinchcombe, 1972, p. 603).

Since the Jencks et al. study—and since Stinchcombe's accolade—other important investigations of IQ heritability have been conducted that go beyond Jencks et al. in complexity and sophistication of models. One such body of literature I consider is the research using path analysis by N. Morton, D. C. Rao, and associates, geneticists at the Population Genetics Laboratory of the University of Hawaii at Honolulu. Representative works are Rao et al., 1974, 1976, and 1978; Rao and Morton, 1974 and 1978; Morton and Rao, 1978; and Morton, 1974. Indeed the use of path analysis to estimate IQ heritability seems to be catching on (e.g., Loehlin, 1978). I also consider (though in somewhat less detail) the research of J. L. Jinks and L. Eaves of the University of Birmingham, England, and D. W. Fulker of the Institute of Psychiatry, London (Eaves, 1975; Fulker, 1975; Jinks and Eaves, 1974; Jinks and Fulker, 1970), as well as the "kinometrics" research of Taubman (1977, which includes Behrman et al., 1977 as well as Jencks and Brown, 1977, which I also consider in due course). In reviewing the Honolulu, the Birmingham, and the kinometrics research I draw upon the lucid and detailed critiques of the econometrician A. Goldberger (especially Goldberger, 1978a, 1978b, 1978c, and 1978d but also Goldberger, 1976a, 1976b, 1976c, and 1977, and Goldberger and Lewontin,

1976). Finally, I examine the methods of heritability calculation and the study of cross-racial adoptions of S. Scarr and R. Weinberg (Scarr-Salapatek, 1971a, 1971b; Scarr and Weinberg, 1975, 1976, 1977, 1978; as well as other works of Scarr and associates).

All of these works, including Jensen's, have something in common: As with any scientific study, they proceed from certain assumptions or premises. Any investigation of any phenomenon necessarily must. What is intriguing about the literature on IQ heritability, however, is its heavy reliance on assumptions that are arbitrary, implausible, or both. This book is about those assumptions. As my analysis will show, a condition pervading the works of interest (though to varying degrees) is that the number of unknown quantities is always in excess of the empirical information needed to find out what the unknowns are. This resultant "underidentification," as it is called, causes the researcher to take certain shortcuts. Although the researcher must necessarily make assumptions about certain unknowns, some researchers are considerably more explicit about their assumptions than others. When a string of flimsy and implausible assumptions is made merely for the sake of cranking out numerical values for the quantitative "heritability" of IQ—despite potentially hazardous policy implications for minorities—a researcher is playing the IQ game. A concise definition might thus be offered: The *IQ game* is the use of assumptions that are implausible as well as arbitrary to arrive at some numerical value for the genetic heritability of human IQ scores on the grounds that no heritability calculations could be made without benefit of such assumptions. That reasoning is of course circular.

A HISTORICAL NOTE

Interest in the statistical study of presumed "inherent" (genetic?) differences in intelligence between racial or ethnic groups is generally attributed to a founder of modern statistics, Sir Francis Galton. In fact, it has been observed (by J. Hirsch, among others) that "there is an unbroken line of intellectual influence from Quetelet [the Belgian astronomer-statistician and early developer of the normal distribution] through Galton and Pearson to modern psychometrics" (1971, p. 99). Among the interesting facts about Galton is that he coined the term *eugenics* ("the science of improving stock . . . to give the more suitable races or strains of blood a better chance of prevailing speedily over the less suitable than they otherwise would have had," Galton, 1883, p. 17). He was, on his mother's side, the cousin of Charles

Darwin. Karl Pearson, inventor of the correlation coefficient, was his student and biographer. Galton was also the first to suspect that the study of twins might carry the secret to solving the nature-nurture puzzle: "Seeking for some new method by which it would be possible to weigh in just scales the effects of Nature and Nurture. . . . The life-history of twins supplies what I wanted" (Galton, 1883, originally stated in 1875). His conclusions are well known today: "There is no escape from the conclusion that nature prevails enormously over nurture" (1883, p. 172). His remarks on race were not especially complimentary to minorities of the day: "The fact of an individual being naturally gifted with high qualities may be due either to his being an exceptionally good specimen of a poor race, or an average specimen of a high one" (p. 198); and "the very foundation and outcome of the human mind is dependent upon race" (p. 217). Yet in places Galton displayed caution, stating for example that "the interaction of nature and circumstance is very close, and it is impossible to separate them with precision" (p. 131).

Why is there so much current and persistent concern over the genetic heritability of IQ? Why is there so much interest in arriving at an actual numerical value for IQ heritability? Why should scientists be intrigued by the prospects of genetically based intellectual differences between races? If history provides any clues, it might suggest the tentative hypothesis that interest in the systematic study of "inherent" intellectual differences between ethnic and racial groups seems to rise whenever a particular minority exerts pressure on the society for what that minority perceives as its just rights and freedoms (see Haller, 1963; Jordan, 1974; Osofsky, 1971).

Thus, for example, the great migration of blacks from South to North in the period 1890–1910 and its associated hints of a new independence and militancy were accompanied in the *scientific* literature of the day by such unflattering characterizations of blacks as "a fungus growth" in need of extermination (Osofsky, 1971, p. 26, quoting several writers such as F. S. Hoffman, W. P. Calhoun, and R. S. Shufeldt). Just prior to the 1920s, the influx of Jewish, Italian, Polish, and Russian immigrants from abroad accompanied the use of the Binet IQ test, newly translated by Lewis Terman of Stanford University and later to be famed as the "Stanford-Binet" IQ test, to demonstrate among other things that "83 per cent" of the newly arrived Jews on Ellis Island were "feeble-minded" (Kamin, 1974, p. 16, quoting Henry Goddard, who administered the IQ tests). In the mid to late 1960s, the civil rights movement and the resultant new black consciousness and ethnic pride, along with demands for cultural recognition and the founding of Afro-American and black studies departments and programs at universities, were closely followed by the appearance of the works of Shuey (1966), Jensen (1967), and of course Jensen (1969).

I intend neither to belabor nor to overdraw the historical relationships between current hereditarian literature and the older hereditarian literature. Even granting the historical connections, it is clear that there are great differences in method, data, and conclusions. My task here is not a historical survey but a detailed methodological analysis. In particular, I propose to take a hard look at such matters as: What assumptions does the researcher make? Are the assumptions explicit or implicit? Would changing an assumption alter the researcher's conclusion a great deal or only slightly? What can we say about the quality of the data used? Are primary sources accurately cited? Is simple arithmetic performed correctly?

Although I do *not* argue that research on either within-group or between-group differences in IQ heritability should be banned, on the basis of what emerges during the course of my review as an inability to answer the quantitative questions being asked, I am not able to encourage such research. I neither argue nor conclude that the heritability of IQ is zero; but I suggest that both the evidence and the methodology indicate that "low" or even "very low" IQ heritability is eminently plausible. Indeed, I conclude that the genetic heritability of IQ is not an empirically estimable quantity, even within a given group or population of individuals. I do not espouse a raw and naive environmentalism, but I am sensitive to the array of environmental variables and effects ignored in the heritability literature. I do not claim to review the literature in the field comprehensively, but I do claim to look critically at a sampling of representative works. I place primary emphasis on methodology and secondary emphasis on data quality. Finally, I note that reasonable people can disagree vigorously on important issues; that serious scientists studying the same subject can differ dramatically in their conclusions; and that a fuller knowledge of the causes of human variation can be attained by focused and vigorous criticism.

Chapter 2

Estimating IQ Heritability

In this chapter I undertake an introductory discussion of the nature of IQ heritability and how it is calculated, focusing primarily on what is called data quality. Later I take up specific methodological issues and assumptions in detail. Here I examine certain aspects of the data and logic pertaining to the analysis of identical twins, fraternal twins, ordinary brothers and sisters, and unrelated persons adopted into the same homes. I discuss Jensen's approaches to data analysis; the work of noted British psychologist Cyril Burt; and the issue of the causes of between-race differences in average IQ score. It is to be noted at the outset that the vast majority of data concerning the estimation of IQ heritability pertains to whites. Few data exist on the heritability of IQ for blacks. My critique in this chapter, as throughout this book, thus applies for the most part to data on whites. Jensen, Eysenck, Shockley, and others as well have nevertheless sought to link three factors: IQ heritability within white populations; IQ heritability within black populations; and the heritability of the average IQ difference between whites and blacks. Inferences about between-race differences have rested (however inappropriately) upon the study of IQ heritability within white populations.

THE COEFFICIENT OF HERITABILITY

The idea of the genetic heritability of a characteristic is a concept borrowed by Jensen from the field of behavioral genetics. The principle seeks to express, in statistical (probabilistic) terms, the degree of connection—in fact, causation—between some genotype, or *genotypic variable* (actually a set of variables), on the one hand, and what is called a phenotype, or *phenotypic variable,* on the other. Generally,

10

the genotypic variable is thought of as an independent variable (or set of variables) while the phenotypic variable is thought of as dependent. The phenotype is considered to be directly measurable. Height, eye color, or the presence or absence of earlobes are examples of phenotypes. Genotypes, in contrast, are thought of as "underlying" gene structures (more technically, chromosome *loci*), which through various biochemical processes, and under given intrauterine and other environmental conditions, cause certain traits, characteristics, or attributes to develop in the individual.

In the analysis of *behavioral* phenotypes such as personality predispositions like aggressiveness, introversion-extroversion, or even IQ score, the genotypic variables—should they exist at all—are neither observable nor directly measurable. Any genes for such traits that might exist have never been isolated and seen under a microscope. Their existence is inferred or deduced. In the language of the behavioral sciences, the genotypic variable is a *construct*, an abstract variable or set of variables, the existence of which is inferred from the study of observable (phenotypic) variables.

The fact that constructs cannot be measured does not mean that they are not useful. Indeed, constructs are important to thinking and research in any scientific discipline. Examples would include the construct of absolute zero in physics or the notion of ideal type in Weberian sociology. Though such constructs are useful (even necessary), they are generally regarded as too abstract to be directly measurable. Whether or not a construct is considered directly measurable depends largely upon the state of affairs in the discipline concerned with its study. *Intelligence genotype* is such a construct. So, for that matter, is *intelligence*.

The inference that a score on an IQ test (an example of a behavioral trait) may be determined by one or more unseen genes arises from the same kind of reasoning that geneticists have applied in the past to human physiological traits. Take height as an example: Geneticists have noted that the closer two people are in biological relatedness, the closer they resemble each other in height. Indeed, the closer they are biologically, the more they tend to "look alike" in literally hundreds of respects. Thus, in the case of *collateral* kinships, identical twins—who are genetically identical—are virtually identical in height. (The fact that identical twins sometimes differ slightly in height is clear proof of at least a small environmental effect on even this physiological trait.) Siblings (brothers and sisters) are also similar in height, but less so than identical twins; cousins are less similar than siblings but more similar than a pair of completely unrelated persons. The same kinds of observations apply to *direct line* kinships: Parents and their children tend to resemble each other in height more than grandparents and their grandchildren, and so on. The consistency of such observations has resulted in the inference by geneticists that height is subject to large (but certainly not complete) genetic determination. Thus, genes for height—the construct—are inferred to exist, even though they are not actually measured or seen.

Such, so the argument goes, is the case with IQ. Jensen and many before him have noted that IQ behaves in much the same way as such physical traits as height. IQ tends to run in families. Stupid parents tend to have stupid children; smart parents tend to have smart children. Identical twins do in fact obtain almost identical scores on IQ tests; siblings tend to get similar scores—to a lesser extent than identical twins but to a greater extent than unrelated persons. And so on for other kinds of kinships. Having thus noted that greater IQ similarity seems to accompany greater biological similarity, Jensen and his predecessors (notably Burt) infer that a set of genes for intelligence must indeed exist; that intelligence is passed from generation to generation through chromosomal combinations and DNA mechanisms, in much the same way as eye color, the presence or absence of earlobes, or the ability to taste phenylthiocarbamide (PTC). Note, however, that IQ and intelligence are by no means one and the same: IQ is an observed empirical score; intelligence is an abstract concept. Whether or not an IQ test validly measures intelligence is an utterly unsettled issue. Nonetheless, the literature I review presumes a close connection between IQ and intelligence.

Genes run in families, but then so do socialization, training, education, and thousands of other environmental properties, a circumstance that lies at the very heart of the present controversy: the question of *how much* of the human differences (variation) in IQ is attributable to biological (genetic) transmission and how much to environmental transmission; how much of the IQ similarity between relatives results from their genetic similarity and how much from their cultural and environmental similarity. Hereditarian writing has inferred that the majority of IQ similarity results from genetic similarity and thus from some particular set of genes (that is, a particular genotype). This entire book is an evaluation of the methodology behind that inference.

Let us assume that one indeed has some valid and reliable way to measure a phenotype. Assume further that one postulates the existence of some genetic variable that is "for" it and that "causes" it. In principle, then, one would expect some connection or correlation between the postulated genotype and the measured phenotype. The coefficient of heritability, symbolized h^2, is such a coefficient. According to some sources (e.g., McClearn and DeFries, 1973, p. 201), the coefficient was introduced by the geneticist J. L. Lush in 1940, although references to such a coefficient appear in the earlier work of Holzinger (Holzinger, 1929; Newman et al., 1937, p. 115). Jensen and behavioral geneticists define heritability as the percentage of the total variance in a given phenotypic variable that is explained by (attributable to; shared with) the variance in any genotypic variable.

In theory, the h^2 coefficient thus represents that portion of the individual differences (the variance) in the phenotypic variable that is attributable to unspecified genetic factors. The h^2 coefficient is thus analogous to measures of explained variance common to all the behavioral and social sciences. Thus, h^2 might be thought of as equivalent to the squared correlation between a hypothetical genotype and a measured phenotype. By definition, it varies between zero and 1.00. The higher the correlation (the closer it is to 1.00), the higher the heritability of the phenotype. The coefficient therefore has a certain interdisciplinary appeal. For example, if $h^2 = .30$, then 30 percent of the total variance in the phenotypic variable is explained by a hypothetical genotypic variable, and consequently 70 percent is explained by everything else (environment; gene-environment interaction; gene-environment correlation; other sources). The phrase "any genotypic variable" appears in the definition for a reason: If the (squared) correlation between any genotype and any phenotype is high (that is, close to 1.00), then by definition the genotype becomes, by implication, the genotype "for" the particular phenotype of interest.

The h^2 coefficient is not a characteristic of a single individual, but of the specific population of individuals being studied (say, white, middle-class Americans). It is thus not generalizable from one population to another. Nor, as I show later, does knowing within-population heritability (say, for whites only or for blacks only) permit any conclusions about between-population differences in the phenotype. Heritability pertains only to the particular population studied at the time of study under the (environmental) conditions then extant, and it is not therefore generalizable through time. For example, even if the h^2 for IQ at a given time were perfect (1.00), this would *not* mean that future changes in the environment could not change IQ. Jensen is quite clear on this in several places (e.g., 1973, p. 366), but he has nevertheless often incorrectly implied (particularly in Jensen, 1969, and even in Jensen, 1978b) that higher heritability of a trait means that it is less subject to change through environmental manipulation. In this connection, Feldman and Lewontin cleverly note that even if a disease has perfect heritability, the cure for the disease is accomplished by manipulation of the environment—by medication, rest, or whatever (1975, p. 1,164). The heritability coefficient itself is estimated from sample data, and thus one needs to consider the degree to which the particular sample is representative of the population being studied.

The dependent (Y) phenotype of interest is IQ. Let the symbol V_Y represent the total *population* variance in IQ. Using Jensen's own formulation (1969, p. 34; 1973, pp. 366–69), which in turn is based in large part on the work of Cyril Burt (especially Burt and Howard, 1956; cf. Burt, 1966, who in turn based his work on that of the pioneer statistician R. A. Fisher), the components of the total variance in human IQ can be thought of in analysis-of-variance fashion, as follows:

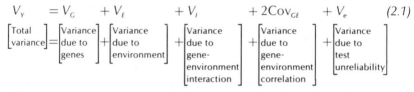

$$V_Y = V_G + V_E + V_I + 2\text{Cov}_{GE} + V_e \qquad (2.1)$$

$$\begin{bmatrix}\text{Total} \\ \text{variance}\end{bmatrix} = \begin{bmatrix}\text{Variance} \\ \text{due to} \\ \text{genes}\end{bmatrix} + \begin{bmatrix}\text{Variance} \\ \text{due to} \\ \text{environment}\end{bmatrix} + \begin{bmatrix}\text{Variance} \\ \text{due to} \\ \text{gene-} \\ \text{environment} \\ \text{interaction}\end{bmatrix} + \begin{bmatrix}\text{Variance} \\ \text{due to} \\ \text{gene-} \\ \text{environment} \\ \text{correlation}\end{bmatrix} + \begin{bmatrix}\text{Variance} \\ \text{due to} \\ \text{test} \\ \text{unreliability}\end{bmatrix}$$

Here V_G represents variance due to hypothetical genetic factors (collectively symbolized as G). V_E represents variance due to *all* environmental variables, including both prenatal (uterine) and postnatal (social, educational, and so forth), all collectively lumped together under E. V_I represents variance due to gene-environment interaction (to particular *combinations* of genotypes and environments). The quantity 2Cov_{GE}, or gene-environment covariance, represents any variance in IQ that is due to any correlation (multicollinearity) between an individual's genes and his environment—as for example when individuals with certain genes presumably select certain aspects of their environment, or when parents provide their offspring with certain kinds of genes (for example, hypothetically, "high intelligence genes") as well as certain ("high" or "enriched") environments. Finally, V_e represents variance due to measurement error—that is, any unreliability of the IQ test being used. (I will show later that contrary to Jensen, V_e contains far more than test unreliability even for this particular equation.) This equation is of course purely theoretical, for how each of these variance components is actually empirically estimated has not yet been specified. Furthermore we have only briefly defined each component here; much more will be said about them later. It is nonetheless a useful equation, stating in the language of analysis of variance (ANOVA) that the total IQ variance is the arithmetic sum of variances due to genotype, environment, genotype-environment interaction, genotype-environment correlation, and measurement error.

Dividing each component by V_Y permits us to express each component as a *proportion* of total variance, such that

$$1 = \frac{V_Y}{V_Y} = \frac{V_G}{V_Y} + \frac{V_E}{V_Y} + \frac{V_I}{V_Y} + \frac{2\text{Cov}_{GE}}{V_Y} + \frac{V_e}{V_Y}$$
$$= (V_G + V_E + V_I + 2\text{Cov}_{GE} + V_e)/V_Y. \qquad (2.2)$$

This in turn permits us to define one kind of heritability coefficient (called *broad heritability*) as

$$h^2 \text{ (broad)} = V_G/V_Y, \qquad (2.3)$$

the proportion of Y's variance attributable to G, or the ratio of genetic variance to the total variance in IQ. (Some discussions define broad heritability to include 2Cov_{GE}.) It therefore follows that

$$1 - h^2 \qquad (2.4)$$

is a measure of all effects other than G that act upon IQ. Thus, $1 - h^2$ is a measure of nongenetic effects, only part of which are attributable to environment as such.

Effects attributable to environment exclusive of other sources can be defined by another coefficient, e^2, called *environmentability*, or *environmental determination* (see Burt and Howard, 1957b; DeFries, 1972; Jencks et al., 1972; Layzer, 1974):

$$e^2 = V_E/V_Y. \qquad (2.5)$$

In the study of IQ, e^2 potentially comprises many factors—effects of socioeconomic status, schooling, cultural advantages, and hundreds of other environmental variables.

V_G itself is sometimes partitioned into its components. Jensen (1969, p. 34; 1973, p. 367), following past researchers such as Burt, employs the following partitioning:

$$V_G = V_A + V_D + V_i + V_{AM}, \qquad (2.6)$$

where V_A is intended to represent *additive* genetic variance, the variance attributable to additive combinations of two or more genes. All *nonadditive* genetic variance is subsumed under the three remaining components: V_D, which represents variance attributable to genetic *dominance* (technically, dominance of one gene over another at the same chromosomal locus); V_I, which represents genetic variance attributable to gene-gene interactions, called *epistasis* (technically, dominance is interaction of gene forms [alleles] at the same chromosomal locus, whereas epistasis is interaction of genes at different chromosomal loci); and V_{AM} which represents variance attributable to *assortative mating*, that is, selective (nonrandom) mating of individuals on the basis of the phenotype (and hence, by presumption, the genotype) in question.[1] Consequently, another kind of heritability coefficient, called *narrow heritability* (after Lush, 1949), can be defined as

$$h^2 \text{ (narrow)} = V_A/V_Y, \qquad (2.7)$$

showing that narrow heritability is the proportion of Y's variance attributable to additive genetic variance and thus excluding any effects of dominance, epistasis, and assortative mating. Broad heritability (equation [2.3]) thus by its definition subsumes all these genetic sources. Dominance, epistasis, and assortative mating are discussed later as the need arises—although the application of notions of dominance and epistasis to intelligence genes seems farfetched, despite attempts to measure dominance through path analysis methods (e.g., Loehlin, 1978).

ESTIMATING IQ HERITABILITY

Theoretically, IQ heritability is the percentage of the total variance in IQ attributable to some unmeasured postulated genotypic variable or set (genotype) of variables. The definition is strictly conceptual. Inasmuch as the postulated genotype is unmeasured, how is IQ heritability empirically estimated? How do Jensen (and Herrnstein, Eysenck, Burt, and others) arrive at an h^2 of approximately .80?

Estimation of h^2 involves two essential strategies. First, it must necessarily be assumed that standard IQ tests indeed measure intelligence with a low degree of error, an assumption accepted for the moment for the sake of argument. Second, if the dependent variable of interest (intelligence) is subject to any genetic determination, the phenotypic similarity reflected by IQ similarity (that is the phenotypic correlation between pairs of individuals) will be high if the biological relationship of the pair is close and low if the biological relationship is less close. According to the hereditarian position, this will persist even if environmental effects are held constant.

For this kind of correlation analysis, which is traceable to Galton himself, the case or unit of analysis is not the single individual but a pair of individuals. The sample (N) consists of N pairs of persons or $2N$ individuals. The two variables consist of a single phenotypic variable measured once on each member of the pair. Given a biologically related pair (twins, siblings, cousins) the IQ score of one member is treated as variable X, and that of the other as Y. Designation of X and Y is purposely arbitrary, but even if X and Y were randomly determined, the size of the resulting Pearsonian correlation, r_{xy}, would be affected. As a correction for this, each pair of individuals can be entered *twice* in calculating the correlation coefficient so that each member of each pair is treated once as X and once as Y, yielding $2N$ pairs and thus $2(2N)$ individuals. If this is done, both X and Y will have the same means and variances. The resulting Pearsonian correlation, called a *double-entry* correlation, is used in many early studies (such as the study of separated twins by Newman et al., 1937, discussed later).

A preferable and presently used procedure for obtaining a phenotypic correlation is the familiar *intraclass correlation coefficient*, r_i, which is defined in standard statistics textbooks as follows (Haggard, 1958; Blalock, 1972, p. 355; cf. Loehlin et al., 1975, p. 287):

$$r_i = \frac{V_B - V_W}{V_B + (n - 1)V_W}.$$

In analysis-of-variance fashion, treating a given pair of individuals as a class such that $n = 2$, the quantity V_W (within-class variance) is calculated in the usual way by summing the squared deviations from the class mean (the mean for the two individuals). The quantity V_B (between-class variance) is thus the sum of the squared deviations of class means from the grand mean, divided by [N(pairs) $- 1$]. Since $n = 2$, the formula reduces to

$$r_i = \frac{V_B - V_W}{V_B + V_W}.$$

The greater the average within-pair difference in IQ (the *less* the within-pair similarity in IQ) the lower the value of r_i. A simple method for testing whether r_i differs significantly from zero is the familiar F-test for ratio of variances, such that $F = V_B/V_W$. Because the double-entry correlation employs all individuals but r_i subtracts one degree of freedom (df) to obtain the between-class variance (V_B), results can differ slightly if the number of pairs is small. In general, however, r_i is virtually identical to the double-entry correlation for a given set of data.

Jensen and the others employ three fundamental methodologies to estimate the magnitude of h^2 for IQ by means of this pair correlation. One involves obtaining the IQ correlation for identical twins who have been raised in separate environments. Another, a collection of methods collectively called *kinship correlation*, involves observing whether or not the IQ correlation for pairs of individuals increases with biological relatedness (genetic similarity). (Included here is the study of parent-offspring correlations, discussed in Chapter 5.) The third involves the study of pairs of biologically unrelated individuals adopted into the same families. In general, the study of separated identical twins is intended to approximate pairs of individuals with identical genes but random (uncorrelated) environments. The study of adopted pairs in the same families is intended to approximate the converse case—pairs with identical environments but random (uncorrelated) genes. The other techniques, involving the study of pairs of varying degrees of kinship, fall between these two.

The Separated Identical Twin Correlation

The most intuitively appealing method of estimating the magnitude of IQ heritability would be to study pairs of individuals who have the closest possible biological relationship but the least similarity in environment. Ideally, this approach would involve the study of monozygotic (one-egg [MZ] or identical) twins separated at the moment of birth and allocated, completely at random, to diverse and uncorrelated environments. It is generally known that monozygotic twins have exactly the same genes. Each is a genetic duplicate of the other, in effect a *clone* of the other. If such pairs were in fact allocated to completely random environments with respect to each other and if the effect of their common prenatal intrauterine environment (twins share a uterus during gestation) were assumed to be negligible, then if the correlation between their IQs were high, the evidence in favor of the hypothesis of genetic determination of IQ would indeed be compelling. Jensen reports that this correlation is about .80. In 1969 he reported it as .75 (p. 49) and in 1970 as .824 in an article devoted entirely to the study of MZ twins raised (presumably) apart (pp. 133, 137). This evidence is quoted extensively by Herrnstein and Eysenck as definitive evidence that IQ is subject to substantial genetic causation. The evidence in favor of genetic determination of IQ thus seems convincing.

Although separation at birth and random allocation to uncorrelated and diverse environments would be an ideal form of experiment, for these twins that is not what happened. The environments of the twins studied by Jensen were in fact markedly correlated. Thus the observed similarity in their IQs could have been caused by the inadvertent similarity in their environments. This issue is of course of the utmost importance in the study of IQ heritability, and I review the evidence of this inadvertent environmental similarity in considerable detail in Chapter 3. In brief, the environments of the twins considered separated by Jensen were actually quite similar because of (1) late separation (many of the twins were separated not at birth, but much later); (2) frequent reunions after their separation but prior to being tested for IQ; (3) upbringings in different branches of the same families rather than in unrelated families; and (4) similarity in educational and socioeconomic environments (a good number of the twins attended the same schools for the same number of years). As I show in Chapter 3, a subsample of truly separated twins (not reunited, not raised in related families, not sent to the same schools) shows an IQ correlation in the neighborhood of .30 or .40. When other factors such as sample bias, measurement error, gene-environment correlation, and the like are also considered, the IQ correlation falls even lower. Such correlations are a long way from Jensen's much-quoted figure of .80.

In order to argue that the IQ correlation of separated MZ twins is an index of IQ heritability, quite a few assumptions must be made: that the twins are separated at birth; that their environments are uncorrelated after separation; that after separation they are placed in a reasonably wide range of different types of environments; that effects of their common intrauterine environment are negligible; that the sample accurately represents the population from which it was drawn. (I examine still other kinds of assumptions in due course.) If in fact all such assumptions are correct, particularly the assumption of uncorrelated environments, the correlation between the IQs of separated MZ twins (r_{MZA}) is indeed an approximation of broad heritability in the population from which the sample of twins was drawn. If all such assumptions are correct, then

$$h^2 = r_{MZA}. \qquad (2.8)$$

Note that the correlation r_{MZA} is itself not squared.[2]

Table 2.1. **IQ Correlations of Monozygotic Twins Reared Apart** (r_{MZA})

Study	Country	IQ test employed (Jensen)[a]	IQ test employed (original authors)	N (pairs) (Jensen)[b]	N (pairs) (original authors)	Correlation (Jensen)[b]	Correlation (original authors)	Correlation corrected for unreliability (Jensen)[b]
Newman et al., 1937	United States	Stanford-Binet	Stanford-Binet	19	19	.77	.67	.81
Shields, 1962	England	Vocabulary test plus Raven's Progressive Matrices	Dominoes plus twice Raven's Mill Hill Vocabulary Test	44	37	.77	.77	.81
Burt, 1966	England	Stanford-Binet	Individual test	53	53	.86	.86	.91
Juel-Nielsen, 1965	Denmark	—	Wechsler-Bellevue	—	12	—	.62[c]	—

[a] Reported in Jensen (1969, p. 52).
[b] Reported in Jensen (1969, p. 52) for Newman et al., Shields, and Burt only.
[c] Reported as .68 in Jensen (1970b, p. 137) because time$_1$ and time$_2$ scores were combined.

Jensen is careful to note that calculation of r_{MZA} is by no means the only way to estimate IQ heritability. Other methods, such as kinship correlation or the study of adopted children, are also used. Jensen, Herrnstein, Eysenck, and many other hereditarians, however, have placed great confidence in r_{MZA}.

To date, there are only four known studies of separated MZ twins. This hinders estimation of h^2 by this method from the outset. Even MZ twins raised together are rare in a general population, and MZ twins raised apart are extraordinarily rare. The studies are necessarily hampered by small sample sizes—thus calling into question the necessary assumption of low sampling bias. It has been possible, however, for researchers to locate and test pairs of MZ twins separated relatively early in life and brought up for a time in separate households for one reason or another—the death or incapacity of a parent, illness, family difficulties, or whatever. The four studies are those of Newman et al. (1937), involving a sample of 19 monozygotic pairs (the only study of separated MZ twins from the United States); Shields (1962), involving 37 pairs from England; Burt (1966), involving 53 English pairs; and Juel-Nielsen (1965), involving 12 pairs from Denmark.

The data from these four studies are summarized in Table 2.1. Results reported by Jensen (1969) for sample sizes, the IQ correlation (r_{MZA}), and type of IQ test used are juxtaposed with those given by the original authors. As shown, there are discrepancies between what the original authors said and what Jensen said they said. These discrepancies deserve comment.

1. Newman et al. (1937) report an r_{MZA} of .67. Jensen (1969, p. 52) reports it as .77. The discrepancy arises because McNemar (1938), in a review of the Newman et al. study, decided to correct the original .67 correlation upward for restriction of range. This correction is based on the contention that the total IQ variance for Newman et al.'s 38 (19 × 2) twins was less than it should be, that is, less than the traditional value of σ^2 or 15^2, the estimated total population IQ variance. (The total sample IQ variance for the Newman et al. twins was 13.00^2.) It can be argued that such a correction is arbitrary and unjustified. It certainly would be if, for example, the restriction in the range of IQ scores was attributable to environmental restriction rather than some form of genotypic restriction. Indeed, it can be argued that if environmental restriction was responsible (certainly a plausible possibility), then the IQ correlation of .67 should be corrected downward. That aside, the value of .77 comes from McNemar, not from Newman et al. As noted by Kamin (1974), the attribution of the .77 correlation to the

original Newman et al. study has been reproduced in articles and text-books over the years. It appears even in textbooks in behavioral ge-netics (for example, McClearn and DeFries, 1973, p. 285). Jensen's er-ror tends in the direction of a hereditarian hypothesis—it leads the reader to believe that the correlation is somewhat higher than that actually reported by Newman et al. In a later publication devoted ex-clusively to the study of separated MZ twins, Jensen correctly reports the correlation as .67, but without explanation (1970b, p. 137). In a reprint of the 1969 article in Jensen's *Genetics and Education* (1972b, p. 127), the correlation also appears as .67—without explanation—although the median value correlation (p. 124), presumably based on the four twin studies, is given as .75, unchanged from the 1969 article.

2. In the first row of the last column of Table 2.1, the correlation of .81 presumably represents Newman et al.'s correlation corrected for attenuation (test unreliability). In general, a correlation like r_{MZA} is cor-rected for attenuation by dividing it by an assumed test-retest reliabil-ity, r_{tt}, for the test in question; that is, by obtaining (r_{MZA}/r_{tt}). This poses two problems. First, the corrected .81 value is the result of correcting the incorrect McNemar correlation of .77 so that from an r_{tt} of .95 for the Stanford-Binet test $(.77/.95) = .81$. The actual corrected value should of course be $(.67/.95)$ or .71. Although Jensen does in fact re-port this value in the reprint of his 1969 article (1972b, p. 127), there is again no accompanying explanation.

The second problem, which is far more general, is of important the-oretical significance: It concerns the advisability of using the attenua-tion correction, a common practice in psychological measurement. Although Jensen employs the attenuation correction in many of his publications and usually reports both corrected and uncorrected figures, he generally bases his conclusions on the corrected figures. The effect of the correction is to inflate the correlation coefficient thus corrected, since it involves dividing by some r_{tt} that is less than 1.00. If in fact such a correction is unjustified on IQ data, then the correction *artificially* inflates the IQ correlation (r_{MZA}) in question and thus also the estimate of IQ heritability, h^2. It can be argued that because of the nonrandom (systematic) measurement error that may underlie IQ tests, corrections for attenuation are unjustified. I return to this issue in Chapter 4.

3. The Shields study is summarized in Table 2.1, row 2. The sample reported by Jensen (1969, p. 52) is 44 pairs. In actuality, although Shields originally sampled 44 pairs, he tested only 38. He himself then rejected 1 of these pairs. (Both twins had been in mental hospitals for

extended periods, and on this basis Shields assumed, perhaps unjustifiably, that their IQ scores were unreliable.) Thus the r_{MZA} of .77 is based on an N of 37 pairs, not 44, as Jensen states. (In later publications, Jensen, 1970b, p. 135; 1973, p. 162, gives an N of 38 for all pairs actually tested.)

4. According to Jensen, Shields's IQ scores were based on a vocabulary test and Raven's Progressive Matrices. This is a gross error. Raven's Matrices, a popular test that Jensen lauds in several places (e.g., Jensen, 1969) as a culture-free test of conceptual ability was not used by Shields. (Nor is the test culture-free, but that is another issue.) Shields actually used a portion of Raven's Mill Hill Vocabulary Test, a test of synonym identifications. His other test was the Dominoes test, an obsolete nonverbal test of logical reasoning developed during World War II. Jensen again corrects himself, this time in 1970b (p. 135), again without explanation.

Shields combined each individual's scores on the tests by adding the raw Dominoes score to twice the raw Mill Hill scores. The resulting scores are thus not IQ scores in the usual sense, as are the scores yielded by the Stanford-Binet or Wechsler inventories which are standardized on large samples; their raw scores are transformed to produce a mean of 100 and a standard deviation of 15. Although Shields did not transform his scores in such a way, Jensen later (1970b) used an unspecified transformation to obtain a mean of 100 and a standard deviation of 15. (See Chapter 3.) Jensen's (1969) assertion that Shields's test consisted partly of Raven's Progressive Matrices thus conveys the false impression that a popular, highly abstract, and presumably culture-free test was used when in fact it was not.

5. For the Burt study (Table 2.1, row 3), the discrepancy concerns which IQ tests Burt in fact administered to his 53 twin pairs. Jensen (1969, p. 52) states that the Stanford-Binet test was used. According to Burt, however, three kinds of tests were administered—a "group test," an "individual test," and something called "final assessments." The correlation of .86 shown in Table 2.1 and reported by Jensen in 1969 is based on Burt's individual test, and the correlations for the group and final assessment tests are .77 and .87, respectively (Burt, 1966, p. 146). Although the individual test may have averaged Stanford-Binet scores with other kinds of test inventories, its exact nature is not described in any of Burt's publications. Nor is it known what tests Burt actually administered to his twins for the other two types. Jensen admitted this somewhat belatedly (1974), after suspicions were raised by Kamin (1973; 1974). (This observation also applies to Burt's data on kinship

correlations, reviewed later.) There is in fact some evidence, cited later in this chapter, that Burt's final assessment IQ scores, and his group and individual scores as well, are not to be trusted. Burt's extensive work on separated MZ twins and on many other kinds of kinship categories, reprinted for nearly 50 years in textbooks and professional articles as definitive evidence concerning the genetic heritability of IQ, is based upon IQ scores that may, in significant part, have been contrived.

6. In a crucial table in his 1969 article (reproduced here as Table 2.2), Jensen lists four studies of MZ twins raised apart. In the text accompanying the table, however, he states that there are "only three" such studies (1969, p. 51). If a fourth study has been omitted from the discussion, it is clearly that of Juel-Nielsen, summarized in Table 2.1, and later reviewed by Jensen (1970). If indeed Jensen did not know of the Juel-Nielsen study or intend to use it in his 1969 article (it is not cited there), the mention of four studies in the table is either a misprint or an attempt on Jensen's part to make his results look more impressive. This error is carried over uncorrected into the reprint of the 1969 article (Jensen, 1972b, pp. 124, 127).

7. In the Juel-Nielsen study, the IQ correlation is .62, the lowest of the four. If Jensen knew of the Juel-Nielsen study when he wrote his 1969 article, he is open to the accusation that he left it out because elimination of the lower correlation makes a hereditarian hypothesis somewhat more impressive. In addition, in his 1970 paper devoted to all four twin studies, Jensen raises the Juel-Nielsen correlation, calculated as an intraclass coefficient, to .68 (1970b, p. 137). This is incorrect. Juel-Nielsen administered his Wechsler IQ test *twice* to each of nine out of his twelve twin pairs, in order to assess test-retest reliability. He then correlated the scores for the first (time$_1$) administration only, arriving at an r_{MZA} of .62. Jensen, however, averaged the time$_1$ and time$_2$ scores for the nine pairs (18 individuals) and correlated them along with the remaining three pairs to arrive at a correlation of .68. When faced with a choice, Jensen evidently selects the larger of two possible correlations—as with his use of McNemar's correlation of .77 rather than Newman et al.'s correlation of .67.

I remind the reader once again that to estimate h^2 using the IQ correlation of pairs of separated MZ twins, one must show (or failing that, simply assume) that the environments of the separated twins are uncorrelated. The lapses in scholarship cited here in fact seem quite minor when one considers just how separated the MZ twins described in these four studies really were. As I show in Chapter 3, the majority of twins in three of the four studies were anything but separated; their environments were markedly correlated and highly similar. The r_{MZA} values given in Table 2.1 are flagrant overestimates of the IQ correlation of MZ twins genuinely separated and reared apart. I postpone detailed consideration of this topic in order to examine other methods used to estimate IQ heritability. These are the various techniques of kinship correlation, including the study of unrelated adopted children raised together.

Kinship Correlation

Because the method of correlating the IQs of separated monozygotic twins is far from perfect as a technique for studying any conceivable effect of unmeasured genes for intelligence upon IQ score, other techniques are also used to see whether h^2 estimated by one procedure (such as r_{MZA}) yields results similar to those obtained by other procedures—to see if they cross-validate each other. One of these techniques, actually a set of related methods, is kinship correlation, developed over a 50-year period (in significant measure through work of Cyril Burt) and used extensively by Jensen and others. In fact, kinship correlation comprises—by a great margin—the bulk of the immense literature on the heritability of IQ.

With this technique the IQ correlation of pairs of persons with a given degree of biological relatedness who were raised together is compared to the IQ correlation for pairs of persons with a different (greater or lesser) degree of biological relatedness who were also raised together. The results of this procedure are then compared to those for individuals of varying degrees of biological relatedness who were raised apart. If the IQ correlation characterizing pairs close in biological relatedness exceeds that of pairs relatively more distant, the evidence suggests genetic determination of IQ.

A common variety of kinship correlation involves the comparison of the IQ correlations of MZ twins raised together (r_{MZT}) with the IQ correlations of dizygotic (DZ) or fraternal twins also raised together (r_{DZT}). A pair of MZ twins will have all of their genes in common, whereas dizygotic or two-egg twins are assumed by geneticists to have, on the average, half their genes in common. In this respect, they are presumed to be genetically no more alike than ordinary siblings. According to genetic theory, r_{MZT} subsumes the effects of both common genes and a common intrapair environment. Similarly, r_{DZT} subsumes the effects of both genes and a common intrapair environment, but the effect of genes on IQ similarity should be less, because DZ twins have fewer genes in common. If one assumes that the degree of intrapair environmental similarity (shared or common environment) is exactly equal for both MZs and DZs raised together and if in fact genes influence IQ to any extent, the difference between the two correlations ($r_{MZT} - r_{DZT}$) is a measure of the effect of genetic similarity upon IQ similarity. If r_{MZT} exceeds r_{DZT}, the difference between the two correlations thus in theory "subtracts out" the effects of environment. If r_{MZT} and r_{DZT} are about the same, the absence of difference is evidence against a hereditarian hypothesis, since it indicates that greater genetic similarity does not accompany greater IQ similarity. Obviously, if r_{MZT} is less than r_{DZT} (especially when the MZs and DZs in question appear in the same study by the same investigator using the same IQ tests), this evidence would completely contradict the hereditarian hypothesis. I show later both that such embarrassing results indeed exist and that it is implausible to assume equal intrapair environmental similarity for MZTs compared to DZTs.

The logic of comparing identical twins to fraternal twins has been applied by Burt, Jensen, and others to correlations characterizing pairs with other degrees of biological relatedness. The implausible assumption of equal intrapair environmental similarity has also been used, although some researchers (e.g., Jencks et al., 1972, and others I consider in due course) rightly question this assumption. In theory, one may compare, in descending order of biological relatedness, for individuals raised together, the following kinship categories: MZ twins, DZ twins and siblings, half-siblings, first cousins, second cousins, and unrelated individuals. By similar logic, parents and their children, grandparents and their grandchildren, and even great-grandparents and great-grandchildren, have been studied. In each case, following a genetic hypothesis and assuming a constant environmental similarity, IQ correlation should decrease directly with biological relatedness.

As intrapair biological similarity increases, however, so does environmental similarity. A biologically similar pair of persons tends to look alike, to interact and associate with one another, and to be treated similarly in their environment—by parents, by other members of their families, by their teachers, and by just about anyone else with whom they come into contact. To the extent that these environmental similarities also produce similarity in IQ, biological relatedness will *appear* to be the cause of the IQ similarity.

In one sense, the inference that genetic similarity produces IQ similarity is correct: If indeed biological similarity results in environmental similarity (that is, if genotype and environment are correlated) and if the environmental similarity in turn produces similarity in IQ, then biological similarity can be said indirectly to "cause" the IQ similarity. But such an inference is meaningless: To conclude that IQ is genetically caused, one must hold constant the extent of environmental similarity, which can be thought of as an intervening variable or set of variables. In comparing any given kinship category to another, one must either demonstrate (or failing that, arbitrarily assume) that both kinship categories are characterized by an equal degree of environmental similarity. For this reason equal environmental similarity must be assumed in order even to attempt an estimate of heritability by means of kinship correlation methods. This assumption is directly analogous to the assumption of uncorrelated social environments in the use of r_{MZA} to estimate h^2. (I stress that I am talking of *social* environment here. To the extent that genes or genetic codes affect some chemical environment [enzymes and amino acids] that might causally affect the development of intelligence, this environment should not be held constant. The causal link of genes to these chemical processes to intelligence to IQ score would justifiably be included under heritability, h^2.)

If similarity in social environment is not assumed equal for any two (or more) kinships being compared, the inference that biological similarity leads to IQ similarity is analogous to the argument that women genetically inherit lower earned income (and lower status jobs) than men because sex is genetically determined! It might be argued with equal persuasiveness that because in some countries women could not inherit land in the Middle Ages, their lack of inherited land was genetically caused, or that because some individuals inherit wealth, their wealth is genetically caused. Such statements are utterly absurd because they do not consider (that is, hold constant) the intervening variables. A familiar rule of all scientific research is thus pertinent: If X causes Z and Z causes Y, to determine whether X causes Y without Z, Z must be held constant.

Two distinct ways of examining kinship data will be considered here. One involves examining the entire array of kinship correlations one at a time; the other consists of systematically comparing kinship categories two at a time; as, for example, MZs versus DZs, MZs versus siblings, and so on. (A third and more contemporary procedure involves attempting to fit a series of simultaneous equations, one for each kinship, to data [the IQ correlations] for a number of kinships using least squares fitting. I review some of these approaches in Chapter 5, especially that of the Honolulu school, Rao et al., 1976, 1978; Rao

Table 2.2. **Jensen's Kinship Table**

Correlations between	Number of studies	Obtained median r^a	Theoretical value[b]	Theoretical value[c]
Unrelated persons				
Children reared apart	4	−.01	.00	.00
Foster parent and child	3	+.20	.00	.00
Children reared together	5	+.24	.00	.00
Collaterals				
Second cousins	1	+.16	+.14	+.063
First cousins	3	+.26	+.18	+.125
Uncle (or aunt) and nephew (or niece)	1	+.34	+.31	+.25
Siblings, reared apart	33	+.47	+.52	+.50
Siblings, reared together	36	+.55	+.52	+.50
Dizygotic twins, different sex	9	+.49	+.50	+.50
Dizygotic twins, same sex	11	+.56	+.54	+.50
Monozygotic twins, reared apart	4	+.75	+1.00	+1.00
Monozygotic twins, reared together	14	+.87	+1.00	+1.00
Direct line				
Grandparent and grandchild	3	+.27	+.31	+.25
Parent (as adult) and child	13	+.50	+.49	+.50
Parent (as child) and child	1	+.56	+.49	+.50

Source: Jensen, 1969, p. 49.

[a] Correlations not corrected for attenuation (unreliability).

[b] Assuming assortative mating and partial dominance.

[c] Assuming random mating and only additive genes, i.e., the simplest possible polygenic model.

and Morton, 1978; and other Honolulu publications. I consider, though in less detail, the Birmingham school, for example, Eaves, 1975; Fulker, 1975, Jinks and Eaves, 1974; and the work of Loehlin, 1978, as well.)

I begin with an overview of the separate kinship categories, using Jensen's much-cited kinship table (Jensen, 1969, p. 49, reprinted here as Table 2.2) as a point of departure. A good part of this book deals with this table and what underlies it. It is based largely upon an equally well known chart published in 1963 by L. Erlenmeyer-Kimling and L. F. Jarvik (Figure 2.1) but includes later data (since 1963) from Burt.

Figure 2.1 appears at first glance to be an impressive compilation, summarizing 99 samples across 52 studies (some studies contain two types of samples); but although the median values indeed increase as biological kinship increases, there is also considerable dispersion around these median values. In fact, given the heterogeneity around these median values, one might justifiably wonder whether any medians (or other measures of central tendency) should be calculated at all.

Jensen lauds the figure as "involving over 30,000 correlational pairings from 8 countries in 4 continents" (1969, p. 48). Similarly, Herrnstein feels that the figure (and Table 2.2) "summarizes more sheer data, over a broader range of conditions, than any other chart in the field of quantitative psychology" (1973, pp. 168–69). Eysenck seems similarly overwhelmed by the figure's apparent vastness (1971, pp. 57–59). In fact, Erlenmeyer-Kimling and Jarvik themselves comment on the "remarkable consistency" in their figure (1963, p. 1,477). As Kamin (1974, p. 75) has noted, the Erlenmeyer-Kimling and Jarvik figure has been reproduced unaltered in countless psychology and genetics textbooks. It has been reproduced in some sociology texts as well (see, for example, CRM, 1973, p. 87). Even texts in behavioral genetics laud the figure as "impressive" and "robust" (for example, Mc-Clearn and DeFries, 1973, p. 287). The argument seems to assume that statistics based on a pooled sample of over 30,000 pairs across 52 studies cannot possibly be incorrect.

But Figure 2.1 is considerably more modest than Jensen, Herrnstein, Eysenck, and many textbooks would lead us to believe. As Erlenmeyer-Kimling and Jarvik themselves clearly indicate in a footnote to their article (p. 1,479), some individuals were used more than once in some of the 52 studies summarized in the figure in order to form pairings. This obviously results in an artificially inflated total N of

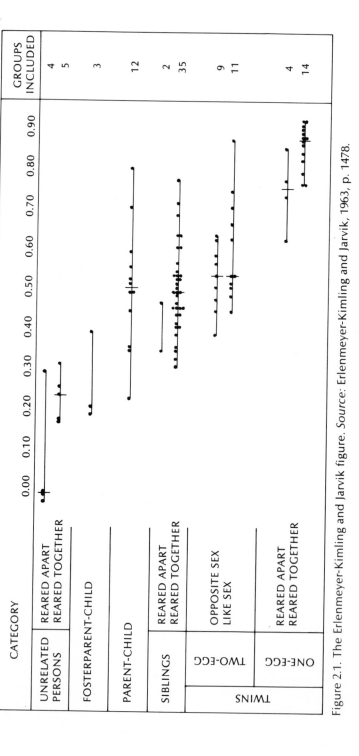

Figure 2.1. The Erlenmeyer-Kimling and Jarvik figure. *Source:* Erlenmeyer-Kimling and Jarvik, 1963, p. 1478.

pairs. If each individual is counted only once and thus allowed membership in only one pair, then according to the Erlenmeyer-Kimling and Jarvik footnote, there are exactly 27,736 pairs—somewhat fewer than "over 30,000"—summarized in the figure.

The actual number of usable pairs is even less than this. Erlenmeyer-Kimling and Jarvik indicate that 15,086 pairs of *unrelated* persons raised *apart* were represented in the four studies. One critic notes that neither genetic nor environmental theory would predict any correlation between the IQs of unrelated persons raised apart (Kamin, 1974, p. 75). Such a correlation would reflect only the unsurprising fact that the IQs of unrelated pairs of individuals selected at random from a general population are not correlated. (The IQ correlation of unrelated persons raised together, a measure of environmental effects, which is quite another matter, is taken up later.) The median correlation of $-.01$, subject to little argument under either hypothesis, is of little, if any, theoretical interest. Subtracting these 15,086 pairs leaves a total N of only 12,650 separate pairs.

The median values themselves appear as vertical lines in Figure 2.1 and (with Burt's results included) in the second column of Table 2.2. The theoretical values in the second and third columns of Table 2.2 were obtained from Burt (1966). Essentially representing the expected proportions of shared genes within a pair based on common parentage, they can be taken to represent the phenotypic correlation expected if only genetic factors caused the phenotypic similarity. For example, siblings and DZ twins have two common parents and are therefore expected to share roughly 50 percent of their genes. A parent-child pair would also share roughly 50 percent of their genes; a grandparent-grandchild pair approximately 25 percent; and so on. Being strictly theoretical, such values have not been empirically observed. The role of such shared-genes values is considered later.[3] In addition, there appear to be some lapses in simple scholarship in both the Erlenmeyer-Kimling and Jarvik figure and Jensen's table. In this regard, Kamin (1973, 1974) has noted important inaccuracies.

Although the list of 52 studies underlying Figure 2.1 has never been published, it can be obtained from Professor Erlenmeyer-Kimling. The list, however, is neither alphabetized nor keyed to the figure; some detective work is needed to find out which data from which studies were included in the figure. In addition, pertinent data from a number of studies included on the list are not included in the figure. Inclusion of these omitted data would change the overall impression of the figure and thus that of Jensen's table as well.

In the category for siblings reared apart, Table 2.2 shows 33 separate studies upon which the median correlation of .47 is based. Figure 2.1, however, indicates only 2 such studies. The source of the discrepancy is an article by Burt on kinship correlations, in which he reports a median correlation of .47 based on "33" studies (1966, p. 150). Jensen later stated that Burt told him privately that this was a "misprint" and that the correct number of studies is 3 (Jensen, 1974, p. 20). In a subsequent reprint of this table (1972, p. 124), Jensen corrected the number to 3 but left the median correlation unchanged. The double error has also been reproduced unaltered by Herrnstein (1973, p. 169) and Eysenck (1971, p. 58), and in textbooks in psychology and other fields.

The discrepancy remains unresolved, however, since the number of sibling-apart studies given by Erlenmeyer-Kimling and Jarvik is two, not three. The two studies included by Erlenmeyer-Kimling and Jarvik are evidently those of Burt and Howard (1956), which gives a sibling-apart correlation of .46, and Freeman et al. (1928, p. 184), which gives a correlation of .34. These values yield a median of .40, not the .47 Jensen copied from Burt. The writings of Burt and Jensen, however, yield no clues about the third study.

Kamin (1974, p. 77) indicates that the most likely candidate as the mysterious third study is Hildreth (1925), which gives a correlation of .23. This study is included on the unpublished Erlenmeyer-Kimling and Jarvik list, but the correlation value is not included as a data point in Figure 2.1. In fact, Erlenmeyer-Kimling and Jarvik evidently erroneously included the Hildreth correlation in their compilation of siblings reared *together*. Taking into account sample sizes, inclusion of this value reduces the median correlation from .40 to .25. It would also reduce the Burt-Jensen median correlation of .47 to .25, a value considerably lower than the theoretical values of .52 and .50 predicted by a strict genetic model. These data in fact suggest that the IQs of siblings reared apart are subject to considerable environmental influence and little genetic influence.

In the category of siblings reared *together* (SIBT), Erlenmeyer-Kimling and Jarvik report 35 studies. There is evidence uncovered by Kamin, however, that the data for this category were compiled in such a way as to bias the median correlation upward. This has two consequences: First, it would make the median r_{SIBT} appear closer to the values predicted by genetic theory; and second, it would reduce the difference between the median correlation for SIBTs and the median correlation for dizygotic twins reared together (DZTs). A strict genetic hypothesis predicts that siblings and DZ twins, being genetically alike, should correlate about the same in their IQs. In contrast, the environmental hypothesis predicts a higher correlation for DZs, who, being the same age, are more likely to be exposed to almost identical environments at home and in school than ordinary siblings. DZTs should thus show a higher IQ correlation than SIBTs.

At least two studies included on the Erlenmeyer-Kimling and Jarvik list contain data that would reduce their r_{SIBT} median to something less than the reported value of .49. (The median value of .55 given by Burt and Jensen would also be lowered.) These studies (Snider, 1955; Stocks and Karn, 1933) are indeed on the list, but pertinent correlations are not included in the figure. Both contain IQ data on the correlation between a single sibling and one sibling chosen randomly from a monozygotic twin pair. Kamin (1974, p. 84), using data from Stocks and Karn, finds the pertinent twin-to-sib correlation for 104 pairings to be .12. The twin-to-sib correlation for 329 pairings in the Snider study is .26. Both correlations, themselves legitimate estimates of r_{SIBT}, are lower than the Erlenmeyer-Kimling and Jarvik median value of .49, the Burt-Jensen value of .55, and the lowest value given on the Erlenmeyer-Kimling and Jarvik figure itself.

Figure 2.1 and Table 2.2 may thus overestimate the population value of r_{SIBT}. Even further evidence of this is contained in at least one study (Pintner, 1918) included in neither the Erlenmeyer-Kimling and Jarvik figure nor their reference list (although a later study by Pintner et al., 1937, does appear on the list). Pintner (1918) reports an IQ correlation of only .22 for 180 pairs of ordinary siblings. It thus appears that a truer population r_{SIBT} could be below .50, although just how far below depends upon what studies one includes. In any case, an r_{SIBT} of less than about .50 would definitely change the orderliness of the Erlenmeyer-Kimling and Jarvik figure.

Other inaccuracies pertaining to discrepancies between Figure 2.1 and Table 2.2 deserve comment. Glancing across all kinship categories listed in Table 2.2 and comparing them to Figure 2.1, one notes that there is agreement in the number of studies reported for the following kinship categories: unrelated children reared apart (4 studies); foster parent and child (3 studies); unrelated children reared together (5 studies); dizygotic twins reared together, different sex (9 studies); dizygotic twins reared together, same sex (11 studies); monozygotic twins reared apart (the 4 studies noted earlier); and monozygotic twins reared together (14 studies). The Juel-Nielsen (1965) separated twins are included in a 1962 paper by Juel-Nielsen and Mogensen as reported in Erlenmeyer-Kimling and Jarvik (1963, p. 1,479). For siblings reared together, Figure 2.1 shows 35 studies, and Table 2.2 shows 36. The extra study is evidently that of Burt (1966, p. 150). Similarly, the extra study given in Table 2.2 for parent (as adult) and child (13 versus 12) is that of Burt (1966), as is the single study for parent (as child) and child. In all other categories, there are puzzling discrepancies—where Jensen has presumably included additional studies. I have already commented on the mythical 33 studies of siblings reared apart and the selective citing underlying the sibling-together category.

The remaining kinship categories are also inadequate. Only one study (Burt, 1966) is given for second cousins; he found a correlation of .16. The only study of uncle (aunt)-to-nephew (niece) correlation is again that of Burt (1966). That these Burt correlations are suspect is discussed in the section immediately following. The same is the case for the one study (Burt's) of the parent (as child)-to-child correlation. There are indeed three studies of first cousins, one by Burt; but it is difficult to ascertain just how the median correlation (.26) was obtained from them (Kamin, 1974, p. 86). While Jensen gives three studies for the grandparent-grandchild correlation, only one (Burt, 1966, with the IQ test unspecified) can be located. Burt's published work, like Jensen's, contains no clues concerning the two remaining grandparent-grandchild studies, and whether they indeed exist is a complete mystery. (Not even fairly extensive investigations such as those of Jencks et al., 1972, or Rao and Morton, 1978, have turned up any evidence as to the whereabouts of these two elusive studies.) It is possible then that some less-than-scientific procedures underlie what otherwise masquerade in the professional literature as sophisticated scientific techniques for studying the genetic heritability of human IQ.

THE CYRIL BURT SCANDAL

Sir Cyril Burt was knighted in 1946, the first psychologist to be so honored. He died in 1971, and serious suspicions about the nature of his published research began with Kamin (1973, 1974) and Jensen himself (1974). These initial suspicions were followed by a front-page news story in the October 24, 1976 London *Sunday Times* by medical correspondent O. Gillie who called the Burt scandal "the most sensational charge of scientific fraud in this century" (reprinted in Gillie, 1977, p. 469). In the News and Comment section of *Science* Wade noted that if accusations against Burt "prove true," the affair "may rank with that of the Piltdown Man in that for years it remained undetected while occupying a pivotal place in a fierce scientific controversy" (1976, p. 916).

Burt is responsible for the largest study of separated MZ twins ever done. It is therefore unfortunate that there are three serious problems with the data he left behind. First, important and detailed descriptive data on the 53 twin pairs are not available because they were evidently destroyed. There is no surviving information about how far apart the twins lived after their separations; whether they attended the same schools; or the extent to which their homes after separation were similar. Jensen tried to obtain the unpublished descriptive data himself in 1974: "In hopes that some of the original data might be recovered after Burt's quite sudden death . . . I corresponded with Burt's personal secretary. . . . But by then, alas, nothing remained of Burt's possessions . . . and boxes of old data . . . were disposed of." All that remained were "the results of the statistical analyses" (pp. 24–25).

The reasons offered for the destruction of Burt's data files vary. According to Wade, Burt's data files were in such disarray that on the advice of L. Hudson of Edinburgh University they were destroyed (1976, p. 917). According to another source, professor of psychology J. Cohen (writing in the *Times* of London, November 10, 1976), Burt was fire-bombed during World War II and forced to move several times after 1940, losing or misplacing data along the way. Cohen also notes that Burt lacked secretarial, clerical, and storage facilities, and found it difficult to locate his own data. Nevertheless, the unavailability of any descriptive data from so eminent a scientist on so large a sample for data so important is bizarre.

Two additional problems with Burt's research affect not only his study of separated twins but certain portions of his data on other kinships as well. In the first place, it is uncertain which tests Burt administered to his sample of separated twins. In the second, throughout Burt's publications from 1955 through 1966, various invariant correlation coefficients are reported, often identical to three decimal places despite the changing composition of the samples over the years.

In the matter of the IQ tests, as already noted, the published work of Burt and his associates refers to three kinds of intelligence tests: a group test, an individual test, and what Burt called a final assessment. As Kamin initially noted (1974, pp. 41–47), even from a close reading of Burt's publications, it is impossible to determine which IQ tests were administered when and to whom. Citing the "group test of intelligence" used over a 45-year period, Burt (1966, p. 140) refers the reader to two previous sources (Burt, 1921, 1933). These sources, however, give no clear indication of what tests were used on which samples.

In the 1966 paper, Burt also refers to the individual test. His description implies that all twins were given some version of the Stanford-Binet scales but ambiguously states only that it was "used primarily for standardization." (In Table 2.1 the reported separated twin correlation for this test is .86.) Burt also indicates that certain "doubtful cases" were given performance tests, but no separate correlation is given for these tests. It is thus not clear whether the .86 correlation includes this performance test as well as the individual test. In fact, in a 1958 paper, Burt states that the .86 correlation was based on "non-verbal tests of the performance type" despite the reference to the Stanford-Binet scales in the 1966 paper.

The third kind of test Burt refers to is his final assessment. He never published the actual scores from these final assessments, which were presumably based on the 53 separated twin pairs. Jensen published them in 1970 as data given to him privately by Burt and calculated an IQ correlation (intraclass) of .88 (1970b, p. 137). Although Jensen called the scores an "English adaptation of the Stanford-Binet" (1970b, p. 135), they were clearly not obtained from any Stanford-Binet. As Kamin has noted (1974, p. 42), a twin identified as Llewellyn in both 1958 and 1966 was reported to have an IQ of 137 as a young man. One of Burt's collaborators elsewhere describes the same man as having the "reading" and "verbal ability" of "a child of barely eleven" (Conway, 1958, p. 186). Imagine an adult with a tested Stanford-Binet IQ of 137 with the verbal abilities of "a child of barely eleven."

No scores exist for any twin pair for any test that constituted even part of the final assessment—they are unavailable (Jensen, 1974, p. 15). Although we have no way of knowing, an extremely disturbing hint about the nature of these final assessment scores is given by Burt and another collaborator, M. Howard (Burt and Howard, 1957a, p. 39; cited by Kamin, 1974, p. 43; and Jensen, 1974, p. 3): The purpose of changing test scores into "assessments" was that *"by these means we can reduce the disturbing effects of environment* to relatively slight proportions" (italics added). This is an extraordinary statement coming from someone skilled in multivariate analysis. How can one partition IQ variance into genetic and environmental components if the "disturbing effects of environment" have been removed beforehand? Although these scores are used in Jensen's analysis of separated twins, they are clearly discredited as scientific data. Their use to ascertain the magnitude of IQ heritability constitutes a peculiarity so great that Kamin was motivated to call it "a rare moment of high comedy in the heritability literature" (1974, p. 43). The uselessness of the scores is certainly not alleviated by Eysenck's effort to explain away Burt's adjustments as deliberate attempts to "produce results nearer to the genotype" (1977, p. 21).

The second disquieting element characterizing Burt's publications is the invariance of correlation coefficients across samples of changing composition, noted initially by Kamin (1973) and then by Jensen (1974, pp. 12–17). In 1955, Burt reported an IQ correlation of .771 based on 21 pairs of separated MZ twins, for the group test. In 1956, Burt and Howard tested the same twins with an unspecified number of pairs added and reported a correlation of .7706 for the group test. In 1958, Burt again reported the correlation as .771 but left the number of pairs unspecified. Finally, in 1966 Burt still reported the correlation for the sample, now grown to 53 pairs, as .771. Assuming that the size of Burt's sample increased from 21 pairs to 53 pairs between 1955 and 1966, the invariance of this statistic considerably strains the laws of probability.

The correlation reported for the individual test is similarly invariant. In 1955, it is .843 (21 pairs); in 1958, it is still .843 (sample size unspecified); and in 1966, it is .863 (based presumably on the final sample of 53 pairs). The final assessment scores reveal a truly amazing consistency: The correlation was .876 in 1955 (21 pairs); .8756 in 1956 (sample size unspecified); .876 in 1958 (sample size unspecified); and .874 in 1966 (53 pairs). Burt's removal of the "disturbing effects of environment" from the final assessment scores seems to have been quite successful.

Similar examples of correlation invariance across samples of changing composition are also present for Burt's data on MZ twins reared together, DZ twins reared together, and siblings reared together. Even Jensen has admitted that such instances of invariance "unduly strain the laws of chance and can only mean error" (1974, p. 24). Later Jensen noted that out of 165 correlation coefficients "with explicit Ns, there are 20 instances where the same r (to three decimal places) is repeated in a later article accompanied by a different N" (1978a, p. 500).

Although the work of Burt, on both separated MZ twins and kinship correlations has stood for some time in the professional literature and in countless texts as definitive evidence on the genetic heritability of IQ, we are clearly forced to have serious misgivings about the usefulness of his data. The 53 "IQ" scores Burt conveyed privately to Jensen in 1970 are not IQ scores: They have been "adjusted" to correlate highly—to "reduce the disturbing effects of environment." To borrow a line from A. Goldberger, Burt's scores may be numbers, but they are not data.

Many have justifiably wondered whether or not the 53 twin pairs ever existed at all. An intriguing question raised by A. Goldberger (personal communication, 1977; see also Goldberger, 1978b, p. F7) is why none of Burt's separated British twins turned up in the Shields (1962) sample of separated British twins. The data-gathering phases of both studies overlapped considerably, and given the extraordinary rarity of identical twins who claim to have been separately raised, as well as the limited geographic area in question and the fact that Shields advertised widely in his search for separated pairs, this does in fact raise some suspicions about the existence of at least some of Burt's pairs. Jensen refers to a personal letter wherein Shields states that he "came across a number of twins who said they had been tested by Burt" (1978a, p. 501). But, and this is a crucial omission, Jensen does not mention that in a letter to him dated June 15, 1973 (a copy of which was made available by Shields himself prior to his own death in the spring of 1979), Shields makes it clear that the few twins he came across who claimed to be Burt's twins were nonseparated MZ pairs. This leaves open the possibility that Burt may have counted some nonseparated pairs among the separated. Even Jensen has become suspicious, wondering "just where, when, and how Burt obtained the test data on the rare 32 identical twins reared apart who were added to his previous twin data between the years 1955 and 1966, when Burt was 73 to 83 years of age" (1978a, p. 502).

Finally, suspicions are certainly not lessened by the conjectures of Gillie (in the *Times* of London, October 24, 1976) and others that Burt's collaborators, Margaret Howard and J. Conway, did not exist. No formal University College or Senate House records of them could be found, nor did "direct inquiries to 18 people who knew Burt" turn up any verification of their existence (Gillie, 1977, p. 470). But one person, Professor J. Cohen, has since stated that he "knew Miss Howard over a period of several years at University College" (the *Times* of London, November 19, 1976). Ms. Conway, however, remains elusive.

Speculation thus abounds about whether the destruction of Burt's data, the vague descriptions of tests used, the invariant correlations, and the unproven existence of the 53 pairs of separated identical twins (as well as that of at least one of Burt's collaborators) constitute simply carelessness or deliberate fraud. Suspicions of fraud loomed considerably larger in the fall of 1978 with the publication of an article in *Science* by Professor Dorfman of the University of Iowa and the subsequent exchange between Dorfman (1979), Stigler (1979), and Rubin (1979). Three of Dorfman's discoveries are of interest: first, the inexplicably close conformity to the normal distribution of the marginal totals in important and widely cited tables from Burt (1961, Tables I and II, p. 11); second, evidence that data in Burt (1961) presented as new and original in fact came from an (uncited) earlier study by Spielman and Burt (1926); and third, Burt's all-too-exact predictions regarding "regression to the mean" of IQ scores.

The suspicious tables cross-classify five occupational categories ("higher professional" to "unskilled") against ten IQ categories (representing IQs of "50 to 60" through "140 +") for adults (Burt's Table I) and their children (Burt's Table II). Both tables show a strong positive relationship between IQ and occupation based on a total "sample" size of 1,000. Burt reports no exact sample sizes, stating only that the marginal totals are "proportional frequencies" per thousand (1961, p. 10), and the "sample" size is thus given as 1,000. These tables have been widely cited to support a genes-IQ-occupation link in numerous articles and textbooks in psychology, sociology, education, and behavioral genetics. They are mentioned by Herrnstein (1973, p. 204), Eysenck (1971, p. 62, although he fails to cite his source), and Jensen (1973, p. 152); even the noted geneticist Dobzhansky (1973, p. 20) describes them.

Dorfman (1978) notes the astonishing fact that the marginal totals for IQ (which are expressed as percentages) for both of Burt's tables conform far too precisely to the theoretical normal distribution to constitute actual data. In a painstakingly detailed analysis, Dorfman (pp. 1181, 1185) fitted 98 frequency distributions from many different studies to the theoretical normal distribution for IQ (33 separate distributions including Army Alpha, Army Beta, and Stanford-Binet scores), height (35 distributions including Quetelet's classic 1870 study of 25,878 American Civil War recruits), and weight (30 distributions). He found, using (χ^2/N) as an index of departure from normality, that Burt's marginals gave the *lowest* χ^2/N (that is, the best fit to normal) of all relative to the other distributions. Dorfman observes that "it may well be that Burt's frequency distributions are the most normally distributed in the history of anthropomorphic measurement. . . . I invite the reader to locate a single unfabricated frequency distribution that gives a smaller χ^2/N than the one obtained for Burt's adult distribution" (Dorfman, 1978, p. 1,181). (The distribution for children was second only to that of Burt's adults.) Such a fit is quite literally too good to be true—particularly considering that according to Burt himself, the study was intended "merely as a pilot inquiry [with] crude and limited data" (1961, p. 9).

Dorfman (1978, pp. 1182–84) also notes that the row (occupational) totals (given as percentages) from Burt's (1961) Tables I and II as well as the virtually equal column totals from his Tables III and IV evidently came from an earlier paper by Spielman and Burt (1926), uncited in Burt (1961). Burt (1961) presented his results as original, citing marginal percentages for occupation (in six categories) of

$$0, \quad 3, \quad 12, \quad 26, \quad 33, \quad 26, \quad -,$$

while the marginal percentages for corresponding occupational categories given in the Spielman and Burt study, which used eight occupational categories, are

$$0, \quad 3, \quad 12, \quad 26, \quad 33, \quad (19 + 7), \quad 0,$$

a close correspondence indeed. This is even more surprising, because Spielman and Burt characterized these percentages as "nothing more than the roughest approximation" (1926).

A third anomaly found in Burt (1961) pertains to the "regression to the mean" phenomenon of IQ scores—a phenomenon also found with other kinds of data. This effect is present if, say, a parent whose IQ is above the population mean of 100 has an offspring whose IQ is slightly less than that of the parent but still above 100; or if a parent with an IQ of less than 100 has an offspring whose IQ is slightly greater than the parent's IQ but still less than 100. Dorfman (1978, p. 1179; 1979, pp. 251–52) notes that Burt's values for the mean IQ of the child ($\bar{X}_c$) for an occupational category is a virtually exact function of the mean IQ of the father ($\bar{X}_f$). Dorfman notes that simply dividing ($\bar{X}_f$ + 100) by 2 gives an exact fit such that

$$\bar{X}_c = \tfrac{1}{2}(\bar{X}_f + 100).$$

This equation tells us that the IQ of an offspring will be exactly halfway between the father's IQ and the population mean of 100. This equation is not a bivariate regression equation; the coefficient ($\tfrac{1}{2}$), which is treated as preset, is not determined by the best fit to the data. When a correlation between $\bar{X}_c$ and $\bar{X}_f$ across occupational categories is calculated (envision a scatterplot of $\bar{X}_c$ values on the $\bar{X}_f$ values), the resulting correlation is $r = .999$! This all-too-close fit caused Dorfman to suspect this equation was a *fabrication device* used by Burt. Collectively, all three anomalies led Dorfman to conclude that "the eminent Briton is shown, beyond a reasonable doubt, to have fabricated data on IQ and social class" (1978, p. 1177).

In the past, Jensen has referred to Burt as "a born nobleman" (1972, quoted in Dorfman, 1978, p. 1177). He has also asserted that Burt's numbers result only from "sheer carelessness" (1978a, p. 500) and from "human frailty found universally in all of us" (p. 502); elsewhere he states that "there as yet has been no evidence of fraud" (1978b, p. 15). He has also noted that "even the most statistically stupid undergraduate could do a neater job of faking his quantitative results, if that were his aim" (1977, p. 471). Other reviewers of the IQ heritability literature, notably J. C. Loehlin, G. Lindzey, and J. N. Spuhler, have asserted that the Burt and Howard (1956) study provides "perhaps the best available evidence on the nature of quantitative inheritance with respect to continuous morphological and behavioral traits in human populations" (Loehlin et al., 1975, p. 91). Such statements about Burt and his work may become considerably more moderate in the future.

JENSEN'S KINSHIP EQUATION VIA JENCKS

Kinship data can also be analyzed by systematic comparison of collateral kinships taken two at a time, where each kinship represents a type of pair raised together. Jensen explicitly devised a formula for this purpose. Although he has systematically applied it only to comparisons of MZ and DZ twins (1967; 1969; 1972b; 1973), Jencks et al. (1972, pp. 283–95) have applied it to siblings and to unrelated adopted persons as well. (I discuss the Jencks et al. analysis of separated MZ twins in Chapter 3 and the analysis of parent-offspring correlations in Chapter 5). Although my evaluation applies more to Jensen's formula than to Jencks et al.'s application of it—indeed, Jencks et al. are aware of some of the equation's inadequacies and place only moderate emphasis on it in their analysis—I also review some inadequacies in Jencks et al.'s analysis of collateral kinships.

Jensen's kinship equation, introduced into the literature in 1967, is as follows (Jensen, 1967, p. 298; 1970a, p. 86; 1973, p. 372; 1976, p. 89):

$$h^2 = \frac{r_{AB} - r_{CD}}{\rho_{AB} - \rho_{CD}} \qquad (2.9)$$

In this generalized kinship equation, which has been used extensively in the research of others, r_{AB} represents the obtained (empirical) IQ correlation between pairs of individuals of a given degree of biological relatedness who were raised together, and r_{CD} represents the IQ correlation between pairs of individuals raised together whose degree of biological relatedness is less than that of pair AB. In the denominator, the correlation ρ_{AB} represents the theoretical proportion of genes shared by individuals making up pair AB, and ρ_{CD} represents this proportion for pair CD. Examples of these theoretical proportions are the Burt-Jensen estimates given in the last two columns of Table 2.2. The greater the difference between r_{AB} and r_{CD}, the greater will be h^2. The logic underlying equation (2.9) is thus quite simple: If in fact the phenotype of interest (IQ) is subject to genetic determination, the phenotypic correlation for a pair with a given degree of biological relatedness will exceed the phenotypic correlation for a pair with a lesser degree of relatedness. The magnitude of this excess is a measure of the magnitude of the genetic heritability of the phenotypic variable.

In order to use this formula to estimate h^2, one must assume that intrapair environmental similarity is exactly equal for pairs AB and CD. One must also assume that the dependence of environmental similarity upon genetic similarity (gene-environment covariance) is equal for both compared kinships. The formula will thus artificially inflate the h^2 estimate to the degree to which the intrapair environmental similarity of pair AB exceeds that of pair CD. The formula also requires that r_{AB} always exceed r_{CD}, although this is not always the case. The Jensen formula extends the logic of an earlier formula for "heritance" devised in 1929 by Holzinger. It is similar, though not identical, to other kinds of heritability estimation formulas appearing in the literature—Falconer's (1960) heritability index, Nichols's (1965) "heritability ratio," and others—all of which I review in Chapter 4, where I also examine in greater detail the algebra and assumptions behind Jensen's formula. The present discussion focuses upon the data to which Jencks et al. apply Jensen's formula.

Jencks et al. applied equation (2.9) to comparisons among monozygotic twins raised together (MZT), dizygotic twins raised together (DZT), siblings raised together (SIBT), and unrelated (adopted) persons raised together (UNT). Equation (2.9) could be extended to include other kinships as well, including half-siblings, first cousins, second cousins, and others. Following both Jensen and Jencks et al., as well as Burt (Table 2.2), it is assumed as a theoretical genetic value that $\rho_{DZ} = .55$, namely, that DZ twins have a bit over half of their genes in common (see Chapter 2, note 3). The same applies to siblings; hence, $\rho_{SIB} = .55$. According to both Jensen and Jencks et al., the theoretical genetic correlation for unrelated persons is expected to be zero; but under an assumption of selective placement of such adopted pairs and any (slight) resultant genotypic correlation, Jencks et al. prefer to set it at $\rho_{UNT} = .03$. Monozygotic twins share all their genes; thus $\rho_{MZ} = 1$. We can now state equation *(2.9)* for almost all possible comparisons in terms of these values:

$$h^2(MZT, DZT) = (r_{MZT} - r_{DZT})/(1 - .55) \qquad (2.10)$$

$$h^2(MZT, SIBT) = (r_{MZT} - r_{SIBT})/(1 - .55) \qquad (2.11)$$

$$h^2(MZT, UNT) = (r_{MZT} - r_{UNT})/(1 - .03) \qquad (2.12)$$

$$h^2(DZT, UNT) = (r_{DZT} - r_{UNT})/(.55 - .03) \qquad (2.13)$$

$$h^2(SIBT, UNT) = (r_{SIBT} - r_{UNT})/(.55 - .03) \qquad (2.14)$$

The sixth possible comparison, that between r_{DZT} and r_{SIBT}, is not stated here. Because siblings and DZ twins do not differ genetically, $\rho_{DZ} = \rho_{SIB}$ and $\rho_{DZ} - \rho_{SIB} = 0$, rendering equation *(2.9)* useless for any such comparison. The comparison of DZs raised together with siblings raised together is nonetheless extremely useful, and inability to extend equation *(2.9)* to this comparison is a flaw in Jensen's invention. It highlights two points to be remembered:

1. Any difference in the IQ correlations of DZTs and SIBTs can reflect only environmental differences (particularly if one assumes no gene-environment correlation). The comparison of DZTs to SIBTs in effect eliminates any effect of genes, since $\rho_{DZ} = \rho_{SIB}$. If in fact $r_{DZT} > r_{SIBT}$, it would strongly suggest greater environmental similarity for DZTs, and thus greater IQ similarity, than that found among SIBTs. The Erlenmeyer-Kimling and Jarvik figure (Figure 2.1) and the Burt-Jensen table (Table 2.2) were evidently constructed so as to underestimate this difference. The Jencks et al. analysis probably also underestimates this difference.

2. If indeed $r_{DZT} > r_{SIBT}$, the assumption of equal intrapair environmental similarity for other kinship comparisons can be called into question: Any two kinship categories of equal genetic similarity can nonetheless differ in intrapair environmental similarity; and by implication, kinships of unequal genetic similarity can also so differ. Demonstration of this would constitute actual empirical evidence that this assumption is unwarranted. Such evidence is reviewed later in connection with the Jencks et al. anal-

ysis. The DZT-SIBT comparison is also discussed later, although this comparison is generally ignored in even the more sophisticated literature (for example, Rao et al., 1976; Rao and Morton, 1978).

A theoretical equivalence between equations *(2.10)*, *(2.12)*, and *(2.8)* deserves mention. The IQ correlation r_{MZT} subsumes the effects of both common genes and common environment. The IQ correlation r_{UNT}, however, subsumes only the effects of the common environment of genetically unrelated persons reared together. Thus the difference $(r_{MZT} - r_{UNT})$, the numerator in equation *(2.12)*, theoretically subtracts the effect of a common environment and is thus a direct measure of h^2 even without the slight correction by the denominator of equation *(2.12)*, which is approximately 1.00. Note also that equation *(2.12)* assumes equal environmental similarity as well as equal gene-environment correlation for MZTs and UNTs. Consequently, according to Jensen, $(r_{MZT} - r_{UNT})$ should give approximately the same result as r_{MZA}, the IQ correlation for MZ twins raised apart—assuming of course that the environments of such twins are uncorrelated. These two quantities, in turn, should equal $[(r_{MZT} - r_{DZT})/(1 - .55)]$—equation *(2.10)*. In general, if all assumptions are in fact correct, the following three-way equivalence should hold:

$$(r_{MZT} - r_{DZT})/(1 - .55) \cong (r_{MZT} - r_{UNT}) \cong r_{MZA} \cong h^2.$$

By these three methods Jensen arrived at his well-known .80 estimate of h^2 for white populations. Ideally, of course, all the kinship equations—*(2.10)* through *(2.14)*—should give approximately equivalent results for h^2, allowing for sampling variability and measurement error across different studies. The comparison of MZTs to DZTs was the subject of Jensen's seminal 1967 article where on the basis of the Erlenmeyer-Kimling and Jarvik compilation, he noted that "the best estimate of h^2 would be obtained from values of $[\rho_{DZ}]$ close to .55 . . . [which] yields $h^2 = .80$" (p. 301). Two years later, he noted that the quantity $(r_{MZT} - r_{UNT})$ is approximately .80 (1969, p. 51). In his 1970 article on separated MZ twins, he notes that their IQ correlation (r_{MZA}) is also approximately .80 (1970b). Faced with such apparently overwhelming consistency, Jensen argued that these three separate ways of estimating h^2 for IQ "cross-validated" one another.

In order to examine the Jencks et al. application, I consider first the studies Jencks et al. used for this purpose. These studies, listed in Table 2.3, deserve extensive comment.

While one might imagine that Jencks et al. averaged the correlation values across studies within kinship categories to obtain overall estimates of the four necessary quantities (r_{MZT}, r_{DZT}, r_{SIBT}, and r_{UNT}), this is not in fact what they did. Indeed, it can be argued that no averages at all should be calculated, given the considerable heterogeneity (not stressed by Jencks et al.) of correlation values across studies within a kinship category. The values for r_{MZT} range from .76 to .92; for r_{DZT}, from .44 to .66; and for r_{SIBT}, from .45 to .67. The heterogeneity for unrelated pairs of persons raised together is most extraordinary (this *is* stressed by Jencks et al.): For r_{UNT-AA} (adopted pairs raised together), the values range from .12 to .65; for r_{UNT-AN} (pairs raised together where one is adopted), from .06 to .38; and for r_{UNT-P} (pooled correlations), from .08 to .50. (I calculated the weighted means in Table 2.3 for purposes of illustration later on.) Jencks and his colleagues included or excluded studies according to various criteria—whether or not a study was done on samples from the United States; the type of IQ test employed; the sampling procedures; the size of the *total* IQ variance. Although these criteria are in themselves methodologically legitimate, they are not at all clearly or consistently applied. Jencks et al. then make three separate kinds of estimates of h^2 from equations *(2.10)* through *(2.14)*: first, using Erlenmeyer-Kimling and Jarvik medians; second, using Burt's kinship correlations; and third, using their own averaging procedures. These results are shown in Tables 2.4 and 2.5.

Although Jencks et al. are to be commended for systematically contrasting three data sets in the calculation of h^2 estimates, their tables contain, in addition to some simple arithmetic errors, what appear to be important instances of selectivity that operate so as to give higher h^2 values (in Table 2.5) than would otherwise result. (Compare the h^2 values in Table 2.5 with those in Table 2.6, which are based on the simple weighted means from Table 2.3.)

Of the 18 studies included in Table 2.3, 13 (over two-thirds) come directly from the Erlenmeyer-Kimling and Jarvik list. More important, except in one instance (Skodak, 1950), the only studies that are not also on the Erlenmeyer-Kimling and Jarvik list are those done since Erlenmeyer-Kimling and Jarvik (1963): those of Nichols (1965), Schoenfeldt (unpublished tables, done since 1968), and the suspect kinship data from the Burt (1966) article.[4] Studies prior to 1963 excluded by Erlenmeyer-Kimling and Jarvik are thus also excluded by Jencks et al. Hence, certain studies that show an r_{SIBT} lower than .50 are not included: those of Pintner (1918), Stocks and Karn (1933), and Snider (1955), which show r_{SIBT} as .22 (180 pairs), .12 (104 pairs), and .26

Table 2.3. **Collateral Kinship Studies Employed by Jencks et Al.**

Study	Country	On unpublished Erlenmeyer-Kimling and Jarvik list?	Test	r_{MZT} (N)	r_{DZT} (N)	r_{SIBT} (N)	r_{UNT-AA} (N)	r_{UNT-AN} (N)	r_{UNT-P} (N)
Holzinger, 1929	United States	No	Binet IQ	.88 (25)	.62 (26)				
Newman et al., 1937	United States	Yes	Stanford-Binet	.89 (50)	.63 (50)				
Nichols, 1965	United States	No	National Merit Scholarship (vocabulary)	.86 (687)	.64 (482)				
Schoenfeldt, unpublished	United States	No	Project talent, "IQ composite"	.85 (335)	.54 (156)				
Blewett, 1954	England	Yes	Thurstone PMA	.76 (26)	.44 (26)				
Burt, 1966	England	No	"Burt's Binet"	.92[a] (95)	.53[a] (127)	.53[b] (264)			.25[a] (136)
Eysenck and Prell, 1951	England	Yes	Wechsler-Bellevue	.89 (25)	.66 (25)				
Herman and Hogben, 1933	England	Yes	Otis Advanced Test	.84 (65)	.49 (234)				
Conrad and Jones, 1940	United States	Yes	Stanford-Binet			.50 (312)			

Table 2.3 (continued)

Study	Country		Test						
Hart, 1924	United States	Yes	Stanford-Binet			.45 (399)			
Madsen, 1924	United States	Yes	Stanford-Binet "and others"			.63 (63)			
McNemar, 1942	United States	Yes	Stanford-Binet			.53 (384)			
Outhit, 1933	United States	Yes	Stanford-Binet and Army Alpha			.67 (63)			
Hildreth, 1925	United States	Yes	Stanford-Binet			.63 (450)			
Burks, 1928	United States	Yes	Stanford-Binet				.23 (21)	—	.23 (21)
Freeman et al., 1928	United States	Yes	Stanford-Binet				.40 (93)	.38 (47)	.34 (140)
Leahy, 1935	United States	Yes	Stanford-Binet				.12 (10)	.06 (25)	.08 (35)
Skodak, 1950	United States	No	Stanford-Binet				.65 (41)	.21 (22)	.50 (63)
Weighted means[c]				.86 (1,308)	.58 (1,126)	.54 (1,935)	.42 (165)	.26 (94)	.34[d] (259)

Source: Compiled from Jencks et al., 1972, pp. 286–92.

[a] Correlations are for Burt's individual test IQ scores (Burt, 1966, p. 146).

[b] Correlation is for Burt's final assessment IQ scores (Burt, 1966, p. 146), not for Burt's individual test, as intended by Jencks et al. (pp. 289, 293).

[c] Weighted by sample sizes (number of pairs, n, for each study) such that r (weighted) $= \Sigma rn/N$.

[d] Excluding Burt's correlation of .25 for 136 UNTs.

Table 2.4.   **Final Correlation Values Employed by Jencks et Al.**

Kinship category	Erlenmeyer-Kimling and Jarvik	"London Binet" (N)	"London Binet" corrected for unreliability	U.S. Binet (N)	U.S. Binet corrected for unreliability	U.S. Binet corrected for unreliability and restriction of range
Adopted-natural pairs reared together (UNT-AN)	—	—	—	.26 (94)	.28	.31
Adopted-adopted pairs reared together (UNT-AA)	—	—	—	.42 (165)	.46	.48
All unrelated children reared together (UNT-P)	.23	.252 (136)	.265	.32 (259)	.35	.38
Siblings reared together (SIBT)	.49	.498 (264)	.524	.52 (1,951)	.57	.59
DZ twins reared together (DZT)	.53	.527 (127)	.555	.63 (50)	.68	.70
MZ twins reared together (MZT)	.87	.918 (95)	.966	.89 (50)	.97	.97

Source: Jencks et al., 1972, p. 293.

Table 2.5.   **Jencks et Al.'s Final Heritability Estimates Using Jensen's Equation** (2.9) **on Three Data Sets**

Type of comparison	Equation used	Estimate using Erlenmeyer-Kimling and Jarvik medians	Estimate using Burt's correlations	Estimate using "Jencks-Moore" means
MZT to DZT	(2.10)	.82	.91	.60
MZT to SIBT	(2.11)	.92	.98	.84
MZT to UNT	(2.12)	.72	.72	.63
DZT to UNT	(2.13)	.63	.60	.62
SIBT to UNT	(2.14)	.54	.50	.40

Source: Jencks et al., 1972, p. 294.

(329 pairs), respectively. Pintner (1918) is not on the Erlenmeyer-Kimling and Jarvik list; Snider (1955) is, but Snider's data for r_{SIBT} are not reported by them; Stocks and Karn (1933) is on the list. The Pintner study was done in the United States and it used the Stanford-Binet. It was excluded by Jencks et al. despite their explicit wish to include American studies employing the Stanford-Binet. The large Snider study is also a United States study; it uses the Iowa Test of Basic Skills (a test believed by some to be more of an achievement test than an ability test). The Stocks and Karn study is English and uses the Stanford-Binet. Exclusion of the Pintner study results in a higher average r_{SIBT} value—making it appear closer to the averaged value for r_{DZT} and to the predicted genetic value (ρ_{SIB}). Exclusion of the Snider and Stocks and Karn studies also inadvertently biases the overall r_{SIBT} upward.

The exclusion of certain studies is evident in another way: Values for only r_{MZT} and r_{DZT} are given for the Herman and Hogben (1933) study although this study also contained IQ data on siblings and found an r_{SIBT} based on 103 pairs of only .32. In principle, it is particularly wise to include correlations from different kinship categories from the same study, thus eliminating problems of using data across different studies with different IQ tests, different time periods, and different countries. Although Jencks et al. seem to agree with this, they exclude the low Herman-Hogben sibling correlation and thus overestimate r_{SIBT}.

Combining the results of r_{SIBT} from the Herman-Hogben, Pintner, Stocks and Karn, and Snider studies produces a weighted mean r_{SIBT} of only .24, considerably lower than the values near .50 reported consistently in the literature. As already noted, it is instructive to compare

r_{DZT} to r_{SIBT}, remembering that any difference between these two correlations can only be attributable to nongenetic factors. If these four studies are averaged with those already included in Table 2.3, the weighted mean for r_{SIBT} becomes .46, slightly below genetic expectations and sufficiently below the averaged value for r_{DZT} to make a difference in heritability calculation. Once again, the literature appears to overestimate the value of r_{SIBT}, thus reducing the (nongenetic) difference between r_{DZT} and r_{SIBT}. Possible errors in judgment begun by Erlenmeyer-Kimling and Jarvik in 1963 are perpetuated in 1972 by Jencks and his associates and thus by recent users of the Jencks et al. compilation as well (including Rao et al., 1976; Rao and Morton, 1978; Loehlin, 1978; Jinks and Eaves, 1974; and Eaves, 1975, all of which I consider in due course).

There are other problems with Table 2.3. From equation (2.10), involving the comparison of identical twins raised together with fraternal twins raised together, the h^2 values can be calculated separately for the 8 studies listed, producing a high h^2 value of $(.92 - .53)/(1 - .55)$ or .87 for Burt and a low h^2 of $(.86 - .64)/(1 - .55)$ or .49 for Nichols. The Nichols study, however, which contains the largest samples of both MZTs and DZTs of all 8 studies, is excluded from the later averaging procedures of Jencks et al., on the grounds that (1) since the National Merit Scholarship Qualifying Test is given to students of above-average ability, their overall variance would be restricted; (2) more DZTs than MZTs were lost during Nichols's sampling procedures; and thus (3) the Nichols value for r_{DZT} of .64 [which Jencks et al. correct for attenuation on the basis of an r_{tt} of only .90 to $(.64/.90)$ or .71], "the highest value reported in the literature for fraternal twins" should be excluded (1972, p. 288).

Jencks et al. must certainly mean literature pertaining to American twins only, for even the most cursory search of the literature shows at least three studies, none of which appears in the Jencks et al. reference list, of MZTs and DZTs showing higher values for r_{DZT} than did Nichols. For example, a study by Husen (1953) in Sweden (reported in McClearn and DeFries, 1973, p. 207, after Vandenberg, 1971, p. 197) obtained an r_{DZT} (uncorrected for reliability) of .70. Husen obtained an r_{MZT} of .90; thus h^2 from equation (2.10) is .44. Wictorin (1952), also in Sweden, obtained an r_{DZT} of .72. Wictorin's $r_{MZT} = .89$; hence $h^2 = .38$. Huntley (1966), in an English study, got an r_{DZT} of .66; his $r_{MZT} = .83$, thus $h^2 = .38$, as with the Wictorin study. Furthermore, the Huntley study, being English, should probably have been included in the Jencks et al. review since Jencks does include four other English studies of MZTs and DZTs (see Table 2.3). It is not clear why Jencks et al. excluded Huntley's study.

Yet another study of MZTs and DZTs done in the United States is excluded entirely from Jencks et al.'s averaging procedures—Vandenberg (1962), which used the Thurstone Primary Mental Abilities (PMA) IQ test. It was excluded on the grounds that another Vandenberg study (Vandenberg, 1967, and thus presumably also Vandenberg, 1962) "does not report simple correlations" (Jencks et al., 1972, p. 285). In an article published before Jencks et al. (1972), Vandenberg himself (1971, p. 197) reports the simple intraclass correlations of .56 for r_{DZT} in his 1962 study (a value close to the average of around .58 for the other studies Jencks et al. use) and a low r_{MZT} correlation of .74—lower than all the other studies of MZTs used by Jencks. This yields an h^2 of only .40 from equation *(2.10)*. Although Vandenberg (1962) is a United States study, the Thurstone PMA scales are used rather than the Stanford-Binet. Jencks et al. are explicit about wishing to use American Stanford-Binet studies, but in their overall review in Table 2.3 they include one study that uses the Thurstone PMA (the Blewett study), and in one place (p. 288) they use the unpublished Schoenfeldt data from the Project Talent "IQ Composite" to calculate an h^2. Their reasons for excluding the Vandenberg study thus remain ambiguous.

Still another study appears in the Jencks et al. bibliography but is excluded from their review—a study by Scarr-Salapatek (1971a) of black and white MZ and DZ twins. Undertaken in the United States it uses the Iowa Test of Basic Skills (which as has already been pointed out may well be an achievement rather than an IQ test). Using the white samples only and pooling Scarr-Salapatek's results across the SES levels she used, one obtains (from pp. 1,291–92) an r_{MZT} of .79; an r_{DZT} (same sex) of .75; and an r_{DZT} (opposite sex) of .69. Pooling the 196 same-sex pairs with the 86 opposite-sex pairs gives an average r_{DZT} of .73, a high value indeed, although, as we noted, Jencks et al. said that the Nichols (1965) value for r_{DZT} of .64 was the "highest value reported in the literature for fraternal twins." Inserting the values of .79 and .73 into equation *(2.10)* gives an extremely low heritability of .13.

For Burt's English study reported in Table 2.3, the IQ test is designated as "Burt's Binet." (By Table 2.4, it has become the "London Binet," as Jencks et al. call it.) This is an error: None of the three kinds of IQ tests discussed by Burt (individual, group, and final assessment), involved a strict Binet test. In any case, the Burt correlations in Table 2.3 involve another error: The correlation values of .92 (for MZTs) and .53 (for DZTs) are those given by Burt for his individual test (1966,

p. 146), yet, the value given for siblings (r_{SIBT}) of .53 is that given by Burt for final assessments (1966, p. 146). The Burt value for the individual test is .498. This is evidently a copying error: Elsewhere (Table 2.4), Jencks et al. use Burt's individual test, since the values of .918, .527, and .498 appearing in Table 2.4 for MZTs, DZTs, and SIBTs, are indeed those given by Burt for his individual test. These errors seem small, however, considering that Burt's IQ scores probably represent simply a bunch of numbers "adjusted" so as to correlate in the way that Burt wanted them to.

The last three columns of Table 2.3 involve unrelated (adopted) persons reared together in the same families. Because such pairs are biologically unrelated, the correlation in their IQs is an estimate of the effect of intrapair environmental similarity upon intrapair IQ similarity (which can arise from adoption agencies' deliberate matching of SES characteristics of adopting and natural parents, that is, through selective placement). Indeed, any correlation in their IQs can be attributable only to environmental effects. It is important to note, however, that this correlation (r_{UNT}) will underestimate the total environmental effect, since it includes only intrapair (family) environment and thus excludes any IQ differences attributable to environmental differences *between* pairs, that is, to specific (nonfamily) environment.[5] It also underestimates overall environmental effects because of restriction of the range in total variance for unrelated persons raised together.

These correlations are instructive. The lowest value is that for the Leahy study (.06); the highest that for the Skodak study (.65) for adopted-adopted pairs—suggesting that individuals having no kinship whatever, even with their adoptive parents, can correlate highly in their IQs. The resulting conclusion that family environmental similarity (a lower-bound estimate of e^2) affects IQ similarity by anywhere from .06 to .65 is clearly less than precise. This lack of agreement is mentioned by Jencks et al., who note that "a large, careful study of unrelated children in the same home is badly needed" (1972, p. 307). Obtaining weighted means (of .42, .26, and .34) is of little help. Indeed, the use of weighted means (or any other averaging procedure) on studies that disagree so vigorously seems wholly unwise. This consideration, however, did not prevent Jencks et al. from using weighted means for their own h^2 estimates (Table 2.5).

Furthermore, Jencks et al.'s weighted mean of .42 for UNT-AA based on 165 cases is in error. On p. 291, Jencks et al. indicate that the Freeman et al. study ($n = 93$ adopted-adopted pairs) includes 21 pairs of "Terman cases"; but Kamin has noted that the 21 Burks cases *are* the "Terman cases" (personal communication, 1979). Thus, Jencks et al. have counted these 21 cases twice in their average correlation of .42 (for 165 cases). The correct weighted correlation is .45 (on 144 cases).

Table 2.4 (Jencks et al.'s Table A-9, 1972, p. 293) shows the correlation values used to calculate the various h^2 values appearing in Table 2.5. Table 2.4 contains two minor arithmetic errors and quite a few major ambiguities.

The first column of Table 2.4 presents the Erlenmeyer-Kimling and Jarvik medians, subject to the biases already discussed. The second column, incorrectly designating Burt's test as a "London Binet," simply presents the correlation values (and sample sizes) given by Burt for his individual test (1966, p. 146). The third column consists of Burt's correlations corrected for unreliability (attenuation); that is, each correlation in the second column is divided by .95, Jencks et al.'s assumed value for test-retest reliability (r_{tt}) for the Binet IQ test. This procedure of course further inflates the correlations. Correcting for unreliability seems wholly unwise here, since it is highly questionable whether Burt used Binet scores. A second problem of major consequence is that the unreliability correction, a popular procedure in IQ testing and educational psychology generally, may not be justified as a general practice at all. It is surely unjustified if the IQ test in question involves systematic rather than random measurement error. Thus far, then, four distinct potential sources of inadvertent bias underlie Jencks et al.'s compilations: exclusion of twin studies that give either a low MZT correlation or a high DZT correlation; possible overestimation of the r_{SIBT} correlation; the use of Burt's untrustworthy test scores and resulting correlations; and the correction for unreliability.

In the fourth column of Table 2.4, there are correlation values of .89 for MZs together (50 pairs) and .63 for DZs together (50 pairs). These are of course the same values listed in Table 2.3 for only the Newman et al. (1937) study. A footnote to Table 2.4 (Jencks et al., 1972, p. 293, n. 4) suggests that these are averaged values, but they are not. The weighted averages in Table 2.3 are somewhat less, at .86 (1,308 pairs) for MZTs and .58 (1,126 pairs) for DZTs. Jencks et al. clearly wish to use only Binet scores, in which case the Newman et al. study is indeed the only usable United States study. All other such studies, which include both MZTs and DZTs listed in Table 2.3, use other tests. It nevertheless seems unwise to ignore *all* these studies for the final compilation in Table 2.4.

In the correlation for siblings in Table 2.4, I observe what in the kindest terms must be called both careless scholarship and bad arithmetic. The sibling correlation is given as .52, based on 1,951 pairs. The six studies of siblings raised together used by Jencks et al. (Table 2.3), all of which employ the United States Stanford-Binet, total only 1,671 pairs. Furthermore, their weighted mean correlation is .54, not .52. Clearly, Jencks et al. used 280 additional pairs from some unnamed study, but no such study appears in any of their tables. It can be neither Burt's English sample of 264 sibling pairs nor the previously ignored Herman-Hogben American sample of 103 pairs. A note to a table appearing in Jencks et al. (1972, p. 289) suggests a seventh study, but it is unlisted and unnamed.

As it turns out, the mysterious study is one conducted by Willoughby (1928, noted by Kamin, personal communication, 1979, and by Jencks, personal communication, 1979). Although Jencks et al. do not cite this study in their bibliography, they used Willoughby's reported r_{SIBT} of .42. There are nevertheless two distinct problems with the Willoughby study: First, it does not use the Stanford-Binet (and by Jencks et al.'s own criteria should not therefore be included in Table 2.4); second, the number of pairs is given in Willoughby as "about 280," not exactly 280. Jencks et al. clearly assumed the sample size was exactly 280. (This would then lead to the weighted correlation based on 1,951 pairs of .52, shown in Table 2.4.)

Finally, the correlation for all unrelated children raised together (r_{UNT-P}) is given as .32 for all 259 pairs (from Table 2.3). This seems to be a rounding error; the weighted mean is .3349. (With Burt's correlation of .25 included, the weighted mean is .3056.)

The final two columns in Table 2.4 involve Jencks et al.'s own corrections for unreliability (fifth column) and for both unreliability and restriction of range (sixth column). Such corrections noticeably inflate all six correlations, especially as compared to the already inflated Erlenmeyer-Kimling and Jarvik median correlations. If the attenuation correction (as well as the correction for restriction of range) is methodologically unjustified, the resulting inflation is artificial. Incidentally, these corrections increase the apparent difference between r_{DZT} (.70) and r_{SIBT} (.59), though not by much.

Finally, Jencks et al.'s own estimates of h^2 use equations *(2.10)* through *(2.14)*, all direct applications of equation *(2.9)*. These are given in Table 2.5. Jencks et al. (1972, p. 295) are justifiably suspicious of these heritability estimates because they use the Jensen equation. The five equations are applied to three data bases: the Erlenmeyer-Kimling and Jarvik medians (from the first column of Table 2.4); Burt's correlations "corrected" for unreliability (and thus further inflated; third column of Table 2.4); and "Jencks-Moore means" (presumably the sixth column of Table 2.4), involving further inflations already noted. There are two arithmetic errors in Table 2.5: First, the h^2 estimate using DZT and UNT for Burt (fourth row, second column) is $[(.555 - .265)/.52]$ or .56, not .60 as Jencks et al. show; second, the h^2 estimate using MZT and UNT for "Jencks-Moore means" (third column, third row) is $[(.97 - .38)/.97]$ or .61, not .63. Both errors thus produced slightly higher h^2 estimates.

More important with respect to Table 2.5 is the fact that the h^2 estimates represent Jencks et al.'s best estimate of h^2 from data on collateral kinships using Jensen's equation. The first observation is that the discrepancies among the fifteen h^2 estimates are extreme. The highest is .98 and the lowest is .40, although these estimates, all direct applications of equation (2.9), should agree reasonably well. They obviously do not. From this mass of evidence, subsuming the work of Erlenmeyer-Kimling and Jarvik, Burt, and the reasoning of Jensen, one is expected to infer that the percentage of variance in (white) human IQ explained by genes "for" intelligence ranges from 98 percent to 40 percent (leaving 60 percent to environment, gene-environment correlation, gene-environment interaction, and a host of other factors). This amounts to saying that IQ is explained simultaneously either mostly by genes or mostly by other factors. That statement is of course meaningless.

The situation is even worse than Jencks et al. suspect. Table 2.6 contains h^2 estimates using equations (2.10) through (2.14) based on the weighted means from Table 2.3, thus including all studies reviewed by Jencks et al. and excluding studies Jencks et al. themselves excluded. Calculations of h^2 are shown separately for weighted means uncorrected for unreliability (column 1) and corrected for unreliability assuming the r_{tt} suggested by Jencks et al. of .92 (column 2). The reliability correction is incorporated as suggested by Jencks et al. (1972, p. 288), that h^2 (corrected) $= [(r_{AB} - r_{CD})/(\rho_{AB} - \rho_{CD})r_{tt}]$. Calculations are shown separately for the three categories of unrelated persons raised together: r_{UNT-AA}, r_{UNT-AN}, and r_{UNT-P}.

The resulting h^2 estimates vary from .71 to .23 (uncorrected) and from .78 to .25 (corrected). They are generally less than those shown in Table 2.5, some by as much as .31. These differences can be accounted for by Jencks et al.'s procedures for inclusion of studies, by their corrections for unreliability incorporated into Table 2.4, and by the fact that Table 2.6 treats the three types of UNT correlations separately.

Table 2.6. **Heritability Estimates from Jensen's Equation** *(2.9)* **Using Weighted Means from Table 2.3**

Type of comparison	Equation used	h^2 estimate uncorrected for unreliability	h^2 estimate corrected for unreliability
MZT to DZT	*(2.10)*	.62	.68
MZT to SIBT	*(2.11)*	.71	.78
MZT to UNT-AA	*(2.12)*	.45	.49
MZT to UNT-AN	*(2.12)*	.62	.67
MZT to UNT-P	*(2.12)*	.54	.58
DZT to UNT-AA	*(2.13)*	.31	.33
DZT to UNT-AN	*(2.13)*	.62	.67
DZT to UNT-P	*(2.13)*	.46	.50
SIBT to UNT-AA	*(2.14)*	.23	.25
SIBT to UNT-AN	*(2.14)*	.54	.58
SIBT to UNT-P	*(2.14)*	.39	.42

The table thus indicates that the heritability of IQ ranges from 78 percent to 23 percent, thus leaving something between 22 percent and 77 percent of the variance to all other factors. Jensen's original cross-validated 80 percent estimate for h^2 therefore appears considerably less than precise. In fact, Jensen says that studies of "monozygotic twins, dizygotic twins, siblings, and unrelated children reared together . . . find values of h^2 in the range from about .60 to .90" (1975, p. 172). From the use here of his own equation this is shown to be untrue.

Because the heritability estimates given in Table 2.6 are based on the weighted means from Table 2.3, they are restricted to the studies that Jencks et al. include in their overall review of collateral kinships. Several studies, as well as data from cited studies, however, were excluded from this table. Table 2.7 shows what the heritability estimates would be if these additional studies and data were included. This table is intended less as a criticism of the Jencks et al. compilation than as an illustration of how greatly h^2 estimates can be affected by the particular studies included or excluded. Table 2.7 averages into the studies reported by Jencks et al. in Table 2.3 the data shown in Table 2.8. For Burt I use the r_{SIBT} based on his individual test as intended by Jencks et al. and not the r_{SIBT} of .53 ($N = 264$) based on the final assessments, and I use $N = 280$ for Willoughby's "about 280" sibling pairs. For the Scarr-Salapatek study, the correlations for same-sex and op-

Table 2.7. **Heritability Estimates from Jensen's Equation** *(2.9)* **Using Weighted Means from Table 2.3 plus Studies from Table 2.8**

Type of comparison	Equation used	h^2 estimate uncorrected for unreliability	h^2 estimate corrected for unreliability
MZT to DZT	*(2.10)*	.53	.58
MZT to SIBT	*(2.11)*	.89	.97
MZT to UNT-AA	*(2.12)*	.44	.48
MZT to UNT-AN	*(2.12)*	.61	.66
MZT to UNT-P	*(2.12)*	.53	.57
DZT to UNT-AA	*(2.13)*	.37	.40
DZT to UNT-AN	*(2.13)*	.67	.73
DZT to UNT-P	*(2.13)*	.52	.56
SIBT to UNT-AA	*(2.14)*	.06	.06
SIBT to UNT-AN	*(2.14)*	.37	.40
SIBT to UNT-P	*(2.14)*	.21	.23

Table 2.8. **Data Excluded from Table 2.6 but Included in Table 2.7**

	r_{MZT}	N	r_{DZT}	N	r_{SIBT}	N
Vandenberg (1962)	.74	45	.56	35		
Scarr-Salapatek (1971a)	.79	78	.73	282		
Burt (1966)					.50	264
Willoughby (1928)					.42	280
Pintner (1918)					.22	180
Stocks and Karn (1933)					.12	104
Snider (1955)					.26	329
Herman and Hogben (1933)					.32	103

posite-sex white DZ twins were averaged to get the r_{DZT} of .73 ($N = 282$) shown in Table 2.8 (combining .75 [$n = 196$] and .69 [$N = 86$] respectively). She reports no sample size for MZTs; following her procedures, I estimate it as 40 percent of her N (of 196) of white same-sex DZTs, or approximately 78 MZT pairs. I exclude the two Swedish studies (Husen, 1953 and Wictorin, 1952) and like Jencks et al. include only studies based on samples from the United States or England.

Averaging these studies in with those in Table 2.3 gives the following overall weighted means (compare these to the bottom row of Table 2.3): $\bar{r}_{MZT} = .85$ ($N = 1,431$); $\bar{r}_{DZT} = .61$ ($N = 1,443$); and $\bar{r}_{SIBT} = .45$

(N = 2,931). The averaged $\bar{r}_{SIBT}$ is noticeably lower than in Table 2.3, which contains only the collateral kinships employed by Jencks et al.

These four reaveraged values were then inserted into appropriate equations *(2.10)* through *(2.14)* to estimate h^2. The heterogeneity of these various h^2 estimates (shown in Table 2.7) is even greater than in Table 2.6. The h^2 estimates uncorrected for unreliability (first column) range from .89 to .06. The h^2 estimates corrected for unreliability (second column) range from .97 to .06. Fourteen of the 22 h^2 estimates are lower than the analogous values in Table 2.6. While one can justifiably argue over which studies should be included in such averaging procedures, Table 2.7 nonetheless gives an idea of what happens when a fairly comprehensive list of collateral kin is used to estimate h^2 from Jensen's equation, thus bringing into question the presumed cross-validation and consistency of the heritability data.

An even greater lack of precision exists in the Jensen kinship equation. Since estimates of h^2 are affected by the value of the sibling correlation r_{SIBT}, only slight differences in its value can result in absurd values of h^2—values greater than 1.00 or less than zero. Using the weighted r_{MZT} means from Table 2.3 in equation *(2.11)*, $(.86 - .41)/.45 = 1$, which means that anything less than a value of .41 for the sibling correlation in equation *(2.11)* will result in h^2 values greater than 1.00. (Recall that our revised weighted mean for r_{SIBT} is .45.)

It is also possible to obtain h^2 values of less than zero, which is absurd. Using the weighted mean from Table 2.3 for unrelated pairs raised together (r_{UNT-P}) of .34, then from equation *(2.14)*, $(.34 - .34)/.52 = 0$, meaning that anything less than .34 for r_{SIBT} in equation *(2.14)* will yield negative h^2 values. If the averaged r_{UNT-AA} ($= .42$) is used, then any value less than .42 for r_{SIBT} in equation *(2.14)* yields negative h^2 values. This shows how sensitive equations *(2.11)* and *(2.14)* are even to small fluctuations in the value of r_{SIBT}. It is important for Burt, Jensen, and the others that the sibling correlation be approximately .50. Otherwise, impossible heritability estimates can result.

The possibility of obtaining h^2 estimates greater than 1.00 or less than zero is not confined to equations *(2.11)* and *(2.14)*, but exists for all five equations. That such absurd results can arise from MZT-DZT comparisons has been noted before (for example, by Sanday, 1972; Kamin, 1974; even Jensen, 1967), but such outcomes can result from other kinship comparisons, as well. Jensen (1967) even goes so far as to imply that the values of the denominators (especially the assumed value of .55 for ρ_{SIBT}) should be changed *in order* to avoid absurd h^2 values. If one could alter parameter quantities in one's model at will in order to force them to conform to one's hypothesized results, science would be a preposterous enterprise, indeed.

Consult Table 2.3. The great variability among studies within a given kinship category can result in quite a few meaningless h^2 values. For equation *(2.11)*, comparing MZTs to siblings, the correlation values for the studies of Burt (.92) and Hart (.45) yield a nonsensical h^2 of 1.04 or 104 percent "explained variance." Using equation *(2.13)*, comparing DZTs to unrelated persons, Blewett (.44) versus Skodak (.65) yields a nonsensical h^2 of −.40, or minus 40 percent, while Eysenck and Prell (.66) versus Leahy (.06) yields an h^2 of 1.15. Thus, comparing different studies can yield h^2 values greater than 1.00 or less than zero even for the same equation. As the reader can clearly see, there are other comparisons among studies in Table 2.3 that yield absurd h^2 values.

Comparisons across studies might not be entirely fair. But one can also discover absurd h^2 values by making comparisons within the same study. In the Scarr-Salapatek study of black and white MZ and DZ twins raised together in two socioeconomic status categories, three kinds of IQ scores were obtained (verbal, nonverbal, and total score) by means of the Iowa Test of Basic Skills—a test of "longterm development of intellectual skills" (1971a, p. 1288). Consequently, 2(race) × 2(SES) × 3(verbal, nonverbal, total) or 12 separate h^2 estimates can be calculated from Scarr-Salapatek's data by equation *(2.10)*. When this is done with her data (1971a, p. 1291), 5 of these 12 h^2 values are negative, since in these 5 instances the MZ correlation was less than the DZ correlation. This is true for both same-sex and opposite-sex DZ twins. For one such comparison (white, low SES, nonverbal score), the h^2 is [(.445 − .619)/.45] or −.39. These negative values are not reported by Scarr-Salapatek, nor are any of Scarr-Salapatek's data used by Jencks et al.

While Jencks et al. do list the Blewett (1954) study in their review (Table 2.3), the MZT-DZT correlations of .76 and .44 are for composite factor IQ scores only on the Thurstone Primary Mental Abilities (PMA) test. If scores on the verbal and reasoning (nonverbal) sections of the PMA are analyzed separately (see Kamin, 1974, p. 103), the resulting MZT and DZT correlations of .73 and .15 for verbal scores yield an h^2 of 1.29. For reasoning scores, the resulting correlations of .71 and .19 yield an h^2 of 1.16. The Stocks and Karn (1933) study obtained MZT and DZT correlations (excluded from Table 2.7) of .84 and .87 (for composite Stanford-Binet scores), thus giving an h^2 by equation *(2.10)* of −.07. Once again, considerable doubt is raised about the cross-validation and consistency of heritability estimates.

A Consistency among the Inconsistencies

Having noted the extreme lack of agreement among the various estimates of h^2 from kinship data, I examine next an interesting consistency that lies beneath the disarray of the h^2 estimates appearing in
Tables 2.5, 2.6, and 2.7. This consistency, unnoted by Jencks et al., can
be accounted for by considering how the social environment operates. The methods of kinship correlation assume equal intrapair environmental similarity for any two kinship categories being compared,
although Jencks et al. do allow for differences in total variance among
kinships. For equation (2.9), for any difference ($r_{AB} - r_{CD}$) to be attributable to genetic similarity, the (unmeasured) intrapair environmental
similarity of pair AB must be assumed to be exactly equal to that of
pair CD, and one must also assume equal gene-environment correlations for any two kinships.

I have already suggested that as genetic similarity between pairs
raised together increases, so does the extent to which they share a
common environment, and thus this common environment could
well account for some, most, or conceivably even all of the measured
similarity in IQ. Consider first the comparison of MZTs to DZTs. It is
certainly plausible to argue that since MZ twins raised together are so
much alike in so many respects, the extent to which they share a common family environment must certainly be greater than the extent to
which DZ twins raised together share a common environment. Indeed there is empirical evidence, ignored by Jensen, Herrnstein, and
Eysenck, but not by Jencks et al., of the relatively greater environmental similarity of MZTs.

For example, an early study by Wilson (1934) showed that MZTs
were more likely to have the same personal friends than were DZTs.
Jones reviews several studies that show that MZTs spend more time
together and are more likely to be in the same classrooms (1954, p.
674). Shields finds that MZTs feel more closely attached to each other
than do DZTs, that DZTs more often actually avoid attachment to one
another (1954, p. 234). An extensive investigation by Koch (1966) finds
that MZTs are more often dressed alike by their parents; they are more
likely to associate with one another, especially in school; and they are
more often placed in the same classroom in school. Scarr (1968) finds
that parents of MZTs tend to treat them more similarly than do parents of DZTs, even when the parents have incorrectly assumed that
true MZs were DZs or vice versa. Scarr has, however, recently questioned these results (Scarr-Salapatek, n.d.). A study by Smith (1965)

finds that MZTs not only are more likely to dress alike, attend sports events together, and have the same close friends, but are far more likely to study together than are DZTs. Certainly, studying together can affect scores on an IQ test. Cohen et al. (1973) find that MZTs are frequently confused with each other by friends and associates, whereas DZTs are rarely so confused. These greater environmental similarities could well account for at least part of the greater observed IQ similarity of MZTs.

Similar reasoning could well apply to other kinds of kinships. The comparison of siblings raised together to DZ twins raised together has already been discussed. Any difference between observed correlations r_{DZT} and r_{SIBT} must be attributable to environmental factors.

The observed difference between r_{SIBT} and r_{UNT} can, however, be attributed to both genetic and environmental sources: The greater IQ similarity of SIBTs could result from their greater genetic similarity or their greater environmental similarity, or both. The same idea of course applies to MZT-DZT (and MZT-SIBT) comparisons—the greater relative IQ similarity of MZTs could result from their greater genetic similarity, their greater environmental similarity, or both. The crucial question is of course how much of the difference in IQ similarity of one kinship compared to another is attributable to genetic and how much to environmental similarity. In Chapters 4 and 5, I consider in some detail the methods used to separate these sources. For the moment, I am concerned only with any effects the heritability estimates in Tables 2.5, 2.6, and 2.7 might inadvertently reveal that could be attributed to differences in environmental similarity.

If indeed the four kinship categories compared are all equal in intrapair (family) environmental similarity, then any h^2 estimates based on equations (2.10) through (2.14) should yield roughly equal results. That they do not leads to the suspicion that differences in environmental similarity across kinship categories might be, at least in part, responsible. This would suggest that h^2 as estimated by the Jensen equation is artificially inflated by the greater relative environmental similarity of one kinship over another.

There are in fact two "pure" comparisons among the five equations (out of ten possible comparisons) which in effect hold genetic similarity constant, and thus provide evidence that environmental differences between kinship categories are indeed inadvertently confounded in Jensen's equation.

Both equation (2.10) (which compares MZTs to DZTs) and equation (2.11) (which compares MZTs to SIBTs) involve comparisons to MZTs.

The denominators in both equations are thus the same $(1 - .55)$ (since DZTs and SIBTs are assumed to be genetically identical), and comparing numerators thus holds genetic similarity constant. The observed excess of the IQ correlation r_{DZT} over the IQ correlation r_{SIBT} can be attributed only to excess environmental similarity of DZTs over SIBTs. Since $r_{DZT} > r_{SIBT}$, it follows on the basis of a strictly environmentalist hypothesis, that $[(r_{MZT} - r_{SIBT})/(1 - .55)] > [(r_{MZT} - r_{DZT})/(1 - .55)]$. This means that h^2 estimated by equation *(2.11)* will be greater than h^2 estimated by equation *(2.10)*. Thus

$$h^2(MZT, SIBT) > h^2(MZT, DZT).$$

If the assumption of equal intrapair environmental similarity were correct, one should not be able to rank these two heritability estimates in this fashion. The fact is, one can. Application of the Jensen equation in this case ignores an elementary algebraic rule that

$$(a - b) - (a - c) = (c - b);$$

the difference between these two h^2 estimates is a simple function of the difference between r_{DZT} and r_{SIBT}. That this difference cannot be genetic is not noted in Jencks et al.'s application of Jensen's equation.

In Table 2.5 the h^2 (MZT, SIBT) based on Erlenmeyer-Kimling and Jarvik medians is .92, while the h^2 (MZT, DZT) based on Erlenmeyer-Kimling and Jarvik medians is .82. For estimates based on Burt's correlations h^2 (MZT, SIBT) is greater than h^2 (MZT, DZT), (.98 > .91), as is also the case for Jencks et al.'s means (.84 > .60). The same holds for the h^2 estimates in Table 2.6 using estimates uncorrected for test unreliability (.71 > .62) and for those in Table 2.6 using corrected estimates (.78 > .68). There are analogous differences in Table 2.7 as well (.89 > .53 and .97 > .58).

Another comparison shows by similar reasoning that Jensen's equation inadvertently includes environmental effects. Equation *(2.13)* (which compares DZTs to UNTs) and equation *(2.14)* (which compares SIBTs to UNTs) both have the same denominators (.55 − .03), and comparing them thus holds genetic similarity constant. It thus follows from a strictly environmentalist hypothesis that

$$h^2(DZT, UNT) > h^2(SIBT, UNT).$$

This is also confirmed in Tables 2.5, 2.6, and 2.7. The appropriate observations from Table 2.5 are that .63 > .54 (for Erlenmeyer-Kimling and Jarvik medians); that .60 > .50 (for Burt's correlation); and that .62 > .40 (for Jencks et al.'s means). In Table 2.6, the prediction holds regardless of whether UNT-AA, UNT-AN, or UNT-P is used. In Table 2.6 (first column), .31 > .23 (for UNT-AA); .62 > .54 (for UNT-AN); and .46 > .39 (for UNT-P). In Table 2.6 (second column), .33 > .25; .67 > .58; and .50 > .42. In Table 2.7 these environmentally produced differences are large: .37 > .06, .67 > .37, and .52 > .21 (first column); and .40 > .06, .73 > .40, and .56 > .23 (second column).

In a logically similar context, Schwartz and Schwartz (1974) have shown that systematic comparisons of kinship categories of equal genetic similarity often produce differences in phenotypic correlation that are sometimes as large as differences presumed by Jensen and others to result from genetic differences. Such nongenetic differences are called *treatment effects*. For example using data from Huntley (1966), Schwartz and Schwartz note that the difference $(r_{DZT(same\ sex)} - r_{DZT(opposite\ sex)}) = .17$, a difference larger than that between the weighted means for r_{SIBT} versus r_{UNT-AA} (Table 2.3) which *does* involve a genetic difference. Schwartz and Schwartz note further that another kind of comparison is instructive, that between male and female DZTs. Using data from Wictorin (1968), $(r_{DZT(male)} - r_{DZT(female)}) = .15$, showing that male DZ twins are more similar environmentally. Using data provided in Jinks and Fulker (1970), and comparing monozygotic twins by sex, $(r_{MZT(male)} - r_{MZT(female)}) = .17$, again a large difference that is environmental. In another paper Schwartz and Schwartz (1975), using data from Vandenberg (1968), find that $(r_{MZT(male)} - r_{MZT(female)}) = .15$, and that $(r_{DZT(male)} - r_{DZT(female)}) = .12$.

It thus appears that certain kinds of heritability estimates based on Jensen's equation can indeed be ranked in accordance with environmentalist predictions, but if Jensen's assumption of equal intrapair environmental similarity across kinship categories were true (at least for DZTs and SIBTs), then heritability estimates from his equation should not be systematically rankable.[6] Other comparisons among equations *(2.10)* through *(2.14)* can be made, but one would not be able convincingly to predict additional rankings on the basis of differences in environmental similarity. Although for these other comparisons, the effects of genetic and environmental similarity could not be disentangled, the results are at least suggestive. For example, an environmental hypothesis would predict $r_{MZT} > r_{DZT}$, but so would a genetic hypothesis. We would thus expect, for example, that $h^2(MZT, UNT) > h^2(SIBT, UNT)$, an expectation confirmed by the data in Tables 2.5, 2.6, and 2.7. Also, while this comparison involves a common element (UNTs), the denominators in the two equations are different, thus not permitting a standardized comparison.[7] According to Jensen's logic, even these heritability estimates should not be systematically rankable. The reader can undoubtedly see other comparisons that yield rankings of h^2 estimates.

A NOTE ON RACE DIFFERENCES AND
BETWEEN-GROUP HERITABILITY

Assume, despite the difficulties just summarized, that the heritability for IQ is a *reliably* (that is, consistently) estimable quantity. In a now classic example, Lewontin in 1970 demonstrated in a few short sentences why one cannot infer the heritability of a between-group mean difference on some Y for two groups (for example, white and black individuals) even if one knows within-group heritabilities for both groups. His example involves two random handfuls of seed of an open-pollinated variety of corn. Each handful will have much genetic heterogeneity in it. One batch of seed is grown in soil to which nutrients are added (the "enriched" environment), and the other in soil with only half the nutrients (the "deprived" environment). After several weeks, the height of each plant (the dependent phenotypic variable, Y) is measured. Within each soil environment, plant heights will differ. This is entirely genetically caused, since no variation in nutrients within a soil environment was allowed. Thus, heritability is perfect, or 1.00. There will, however, be a considerable difference in *average* plant height between the two soil environments, the plants in the enriched environment being taller. Hence the within-group differences are entirely genetic and within-group heritability is perfect; yet the between-group difference in height is entirely environmental; the heritability of the average difference between groups is zero.

Could one thus infer that the 15-point difference in IQ between whites and blacks is genetically caused even if the h^2 for IQ for whites and blacks were known and high? The answer is no. Is there nonetheless some way to express between-group heritability as a function of within-group heritability? The answer is a heavily qualified, strictly theoretical, yes.

In 1972, J. C. DeFries, a behavioral geneticist, stated an equation that expresses the heritability of the mean group difference in Y as a function of within-group heritability. This is what encouraged Jensen to guess that a sizable amount ("between one-half and three-fourths") of the black-white difference in IQ was genetically caused (1973, p. 363). Letting one group be whites and the other blacks (or any other ethnic minority), and assuming a large sample of individuals in each group, the equation is as follows (after DeFries, 1972, p. 9; McClearn and DeFries, 1973, p. 299; first cited by Jensen in a footnote in 1972b, p. 30; and discussed and extended somewhat in Loehlin et al., 1975, pp. 290–91):

$$h_B^2 \cong h_W^2 \; \frac{(1 - t)r}{(1 - r)t} \qquad (2.15)$$

where

h_W^2 = average heritability of Y within groups;
h_B^2 = heritability of the group average (mean); or, between-group heritability;
t = the intraclass correlation for race and Y (observed);
r = the intraclass correlation for race and genotypic variable, G (unobserved).

Note that r is conceptualized here as the intraclass correlation between intelligence genotype (G) and race (group), and is of course unobservable. The correlation between race and IQ (Y) is t; to the extent that biological race is measurable, this correlation *is* observable. The G-to-Y connection is h_W^2 (averaged together for whites and blacks). If one knows t (which we do) and h_W^2 (which the hereditarian literature treats as knowable), then to find h_B^2 all one needs to know is r. In an illustrative figure, McClearn and DeFries (1973, p. 300; cf. De-Fries, 1972, p. 11) plot h_B^2 as a function of h_W^2 for a value of $t = .20$ (based on the observed 15-point IQ difference between blacks and whites) for various assumed values of r (from zero to 1.00). Equation *(2.15)* and the McClearn and DeFries figure show, for example, that if $t = r = .20$, then $h_B^2 = h_W^2$ exactly. It also shows that even if h_W^2 is high (say, .80), then if r is low, say .06, h_B^2 can be only .20 or less—a point made clear by DeFries himself. The interested reader can fit other hypothetical values of r into this equation while keeping $t = .20$ to see how h_B^2 varies as h_W^2 varies. If the h_W^2 even for whites is not estimable (and the conclusion of this book is that it is not), then h_B^2 would not be estimable from h_W^2 even if r were estimable.

But even assuming that h_W^2 is known, then as noted by some (e.g., Feldman and Lewontin, 1975) this equation is useless regarding black-white IQ differences: To get h_B^2, r must be known, but if r were known, we would know what we wanted to know all along anyway and finding h_B^2 would be trivial. (Remember that r expresses the connection between race and *genotypic* "intelligence," G.) Feldman and Lewontin note also that the equation is incorrect in any causal sense, since r is dependent upon h_B^2 and not vice versa (1975, p. 1167). Thus the suggestion that the size of h_W^2 yields information about the size of h_B^2 is incorrect. Furthermore, as McClearn and DeFries (1973) themselves note, the equation applies only under an assumption of zero gene-environment covariance—an assumption contrary to the results of Jencks et al. (1972) who find gene-environment covariance to be quite noticeable for white populations (20 percent). In any case the equation would apply only under an assumption of zero gene-environment interaction, and only under an assumption of zero interaction between race and G, and between race and environment. Even treating h_W^2 as known, there is still only one equation with two unknowns (h_B^2 and r), neither of which can be estimated without knowing the other.

As noted in Loehlin et al. (1975, p. 291), even though h_B^2 may be algebraically expressed without r on the right-hand side, as

$$h_B^2 = \frac{h_P^2 - (1 - t)h_W^2}{t}, \qquad (2.16)$$

h_B^2 is still not estimable since h_P^2 (the total population heritability for all individuals, black, white, and others) is nonestimable (h_W^2 and h_P^2 are not the same quantity). To get this particular h_P^2, as Loehlin et al. note, one would have to study not only black MZ twins and white MZ twins but MZ twins where one was black and the other white—a genetic impossibility. Thus equation *(2.16)* seems no more useful than equation *(2.15)* for inferring h_B^2 from h_W^2.[8]

If one persists in trying to estimate h_W^2 (for blacks as well as whites) and t, are any conclusions possible? To estimate h_W^2 for blacks, black MZ twins are compared to black DZ twins, and using the same IQ test, white MZ twins are compared to white DZ twins to get h_W^2 for whites. To date only four studies have attempted this, and they are utterly inconclusive.

One by Nichols (1970) (cited in Loehlin et al., 1975, p. 109) showed an h_W^2 of .64 for whites (using equation [2.10] on 60 MZT pairs and 97 DZT pairs) using the Bayley Mental Scale IQ test, and an h_W^2 of .68 for blacks (based on 85 MZT pairs and 117 DZT pairs). The white h_W^2 for the Stanford-Binet IQ test was only .23 (for 36 MZTs and 65 DZTs) but .53 for blacks (on 60 MZTs and 84 DZTs), suggesting that the black h_W^2 may be somewhat higher. The Scarr-Salapatek (1971) study showed lower heritability values for blacks based on MZT-DZT comparisons, but this study also obtained negative heritability values. More recent data of Scarr-Salapatek (1974, p. 464) show black h_w^2 across five tests (.59, .28, .42, .61, and .48) to be consistently less than the white h_w^2 values for the same five tests (.87, .49, .58, .71, and .63), evidence that black within-group heritability is lower than white. A single study reported in three papers (Osborn and Gregor, 1968; Osborn and Miele, 1969; Vandenberg, 1970, all reviewed in Loehlin et al., 1975, pp. 103–7) is inconclusive; higher heritabilities are shown for whites (Vandenberg) but also blacks (Osborn and Miele) "from the same data, indeed apparently from copies of the same computer printout" (Loehlin et al., p. 104)!

What can one say about the relationship between race and IQ? According to Jensen, over the last 50 years a large number of studies have shown an average 15-point IQ difference between black and white populations. But two other approaches to this question are of interest. One has been to measure "ancestry" or "racial admixture" (by self-reports of the race of one's direct-line kin) and see if there is a relationship between ancestry and IQ. The other has been to measure blood-group gene "markers" via individual blood type, and then see if this relates to IQ. I will review here a representative sample of studies of both type.

A racial admixture study in Canada by Tanser (1941) showed "the racially mixed group was superior to the Negro [group] but when a more refined breakdown (full, $\frac{3}{4}$, $\frac{1}{2}$, $\frac{1}{4}$) was employed, on two of these tests the group judged to be $\frac{1}{2}$ Negro averaged *lower* than the $\frac{3}{4}$ Negro goup" (cited in Loehlin et al., 1975, p. 121, italics added). A widely cited study by DeLemos (1969), summarized by Eysenck (1971, pp. 95–96) and regarded by Jensen as an "impressive study" (1973, p. 315), reported that Australian aboriginal children who were several generations removed from white European ancestors scored higher on four out of six Piagetian perceptual constancy tests (a presumed measure of intelligence) than did fully aboriginal children, even though the children were reported as being the same in appearance (including skin color) and attended the same school. But as Kamin (1975) and Loehlin et al. (1975, p. 285) have noted, an attempt by Dasen (1972) to replicate these findings in the same Australian community with the same tests failed to show any ancestral differences at all in test score.

Studies have shown little IQ difference based on African blood-group differences although some studies have noted a slight tendency for blacks scored as higher in African blood-group ancestry to score higher in IQ than blacks with less African ancestry. Loehlin et al.

(1973, cited in Loehlin et al., 1975, p. 124) in a study measuring 16 blood-group genetic markers found no overall differences, but did find a nonsignificant difference showing lower scores for blacks with *European* ancestry. In a detailed study done by Scarr et al. (1977) in Philadelphia, odds coefficients for African ancestry were calculated for each individual in black and white samples, based on estimated probabilities of appearance of 12 blood-group marker types from 8 African regions. Certain blood types ("AB" and "B" in the Duffy marker system) showed large differences between black and white samples, thus offering evidence of the validity of such measures as measures of "race." However, only very low and nonsignificant correlations were found between these measures of African ancestry and 6 IQ measures (5 tests plus a principal factor measure). Indeed, her Tables 7 and 8 show 4 (out of 12) correlations suggestive of higher IQ scores for those with the higher odds of having African ancestry. As Scarr et al. appropriately note (p. 85), both findings stand in stark contrast to Jensen's (1973) predictions about racial ancestry and IQ and Eysenck's (1971) speculations.

Such results cause one to conclude that even t in equation *(2.15)* is not as easily estimated as is generally believed. For if no relationship exists between IQ and some biological measure of race (such as blood-group marker), then inferences about race differences in "intelligence" become impossible no matter what the status of within- or between-group heritability.

GENE-ENVIRONMENT COVARIANCE AND INTERACTION

I have argued that any genetic similarity between a given pair of individuals raised in the same family produces similarity in their environments, and this in turn can result in similarity in their IQs. The proportion of the total IQ variance that is explained by this gene-environment relationship falls under the heading of gene-environment covariance. Covariance is theoretically a separate source of IQ variance, as in equations *(2.1)* and *(2.2)*, and is treated here as a separate source of variance unless otherwise noted.

Intrapair environmental similarity can arise from genetic similarity, or from other sources, such as being raised in the same family. This is what is meant by IQ variance that is attributable to environment. (Environmental similarity that does not depend upon genetic similarity can be clearly envisioned in the case of unrelated adopted-adopted pairs raised together since any environmental similarity characterizing them cannot arise from genetic similarity.)

The first source pertains to gene-environment correlation. Assume that some G "for" IQ indeed exists. In a conceptually useful discussion, Plomin et al. (1977) hypothesize at least three ways in which a positive correlation between the genes "for" IQ (G) and environment (E) can arise:

1. *Passive* gene-environment correlation can arise when parents with, say "high intelligence genes" transmit both these genes as well as cultural, educational, and other environmental advantages to their offspring. (This gives the offspring a "double advantage," as Jencks et al. have called it.) Conversely, parents less well endowed genetically may transmit fewer environmental advantages to their offspring—and presumably "low IQ genes" as well.

2. *Active* gene-environment correlation arises if "high G" individuals actively seek out or select from their environments various advantages (as when bright children initiate interaction with other bright children or adults, or read advanced books), while the relatively "lower G" individuals do not select such advantages.

3. *Reactive* gene-environment correlation will arise if persons in the environment react differently to persons on the basis of their apparent brightness—teachers encouraging bright children and discouraging dull ones, for example. It is also conceivable for a negative correlation to arise between G and E, as where the (presumably) "low G" individuals are put into special classes for the retarded with the aim of increasing their phenotype (IQ); or, conversely, where gifted (presumed "high G") individuals are treated as "odd" and thus discouraged by society from showing off too much of their attractive genes.

The prevailing literature conceptualizes gene-environment correlation (covariance) as a correlation between E (all environmental variables taken as a set) and the unmeasured G "for" IQ. But another kind of covariance is inadvertently being included: Clearly, environmental similarity within pairs of individuals raised together (MZ twins, DZ twins, and siblings, but not unrelated individuals) can arise from common genes other than those "for" intelligence. Intrapair environmental similarity can arise from genes "for" physical appearance, height, stature, perhaps even beauty or ugliness, or even certain personality traits, and a long list of physical traits that are unarguably subject to at least some genetic determination—that is genotypes other than our G. People raised together in the same home who look alike also get treated alike in their home, school, or occupational environments. (This is a type of reactive gene-environment correlation.)

Any environmental similarity arising from genes that are shared but that are for, say physical attributes, is thus not covariance between G and E, but covariance between some other genotypes (call them collectively G') and environment. These other genotypes may well produce environmental similarity, and thus IQ similarity, for pairs of greater biological relatedness. This means there are at least two distinct kinds of covariance that can artificially inflate h^2 estimates based on kinship methods or the study of separated twins: correlation of E (environmental similarity) and G (presumed similarity on genotypes for intelligence); and correlation of E and G' (genetic similarity with regard to genotypes other than G). In fact, if one assumes some non-zero correlation between G and G', it can be shown that there are three distinct covariances—between G and E; between G and G'; and between G' and E.[9] And if G does not exist, only the G', E covariance inflates the h^2 estimate—and this covariance is *not* the "gene-environment" covariance of the heritability literature.

The distinction between these two kinds of covariance is not made even in the painstakingly detailed analyses of Jencks et al., or the pioneering work of Sewall Wright (1931; Morton, 1972, p. 256), or the sophisticated research of Rao et al. (1974, 1976) and Rao and Morton (1978). Jencks et al.'s estimate of the proportion of the total variance in IQ explained by gene-environment covariance is about .20, Wright's is about .13, Rao et al.'s about .10 for children and, like Wright, about .13 for parents (1976, p. 238). Jencks et al. talk explicitly about the G, E covariance, not about G', E covariance, although they may subsume the former in the latter. Still another difficulty with these analyses, especially Jensen's, is that the amount of covariance is implicitly assumed to be equal across all kinship categories.

This distinction between the two kinds of gene-environment covariance seems important to keep in mind for the following reasons: First, both kinds constitute potential sources of artificial inflation of h^2 estimates based on either r_{MZA} or kinship procedures. Second, both constitute unknown components of the basic equations for estimating IQ heritability. This latter point is no small matter: As the number of unknown and empirically inestimable quantities in the relevant equations increases, so too does the indeterminacy of IQ heritability itself.

A problem generally considered along with gene-environment covariance is gene-environment *interaction* (discussed in detail in Chapter 4). Interaction is said to exist if the direction (positive or negative) of the relationship between two variables is dependent upon the value of a third variable. Thus if the direction (slope) of the relationship (regression) of IQ on G is different (that is, nonparallel) for different environments; or, equivalently, if the slopes of IQ on E are different for different genotypes, then gene-environment interaction exists. In this case, the relationship between genotype and IQ is conditional upon environment (or, equivalently, the relationship between environment and IQ is conditional upon genotype). If the slopes of IQ on G (or IQ on E) are parallel for different values of E (or of G), then noninteraction or *additivity* exists.

The problem is of course that one never actually "sees" these slopes in the nonexperimental analysis of continuous behavioral phenotypic variables. (In laboratory experiments with animals, however, it is possible through several generations of selective breeding to measure genotype for some phenotypic trait in different environments and thus measure the pertinent slopes.) Interaction must thus be tested for indirectly. Several techniques (reviewed later) have been suggested (for example, Jinks and Fulker, 1970; Plomin et al., 1977). One approach is to see if different h^2 values (where the unsquared h is taken as a measure of the regression, or slope, of IQ on G) are obtained for individuals in different environments. Using this strategy, on the clearly plausible argument that blacks and whites generally experience different environments, especially in regard to socioeconomic status variables, then different h^2 values for blacks and whites would presumably be evidence of interaction. As already noted, the evidence pertaining to different h^2 values for different races is highly inconclusive, but it is worthwhile to point out that Dobzhansky (1973, p. 23) has interpreted Scarr-Salapatek's (1971a) results (cf. Scarr-Salapatek, 1974), which did show different h^2 values for blacks and whites, as evidence of gene-environment interaction.

But this poses a conceptual problem analogous to the dual character of gene-environment covariance.[10] Dobzhansky interpreted different h^2 values for whites and blacks as different regressions (slopes) of IQ on the G "for" IQ for blacks and whites separately (different Es). But since race is genetically caused, an alternative explanation of differing h^2 values for blacks and whites might be that black skin (caused by genes "for" skin color, in other words by G', not G) in turn elicits some set of environmental treatments and thus certain IQ scores, while white skin elicits different environmental treatments and different IQ scores. Perhaps black skin produces a different range (greater or lesser) of environmental treatments from schoolteachers than white skin, thus restricting the range of environments in one group compared to the other, consequently lessening e^2 and possibly increasing h^2. Different resulting h^2 values would thus erroneously be attributed to different G to IQ relationships, and not to differences in the relationships between the genes for skin color and IQ. The whole issue is worthy of more thought than it has been given in the heritability literature.

SUMMARY

In this chapter, I have reviewed three major techniques used by Jensen, Herrnstein, Eysenck, and others to attempt to estimate the genetic heritability of IQ scores: the study of separated identical twins; kinship correlations; and subsumed under the latter, the study of biologically unrelated pairs of persons adopted into the same home. I noted a number of important errors in Jensen's reporting of the original data and results from the four studies of separated identical twins, which inflated the IQ correlation (r_{MZA}) for these twins. Regarding kinship correlations (including the study of adopted pairs), I reviewed the separate kinship categories and the systematic comparison of kinship categories taken two at a time, as demanded by Jensen's kinship equation *(2.9)*. I noted that Jensen's compilation of data (Table 2.2), with its heavy reliance on the Erlenmeyer-Kimling and Jarvik figure (Figure 2.1) and the untrustworthy correlations reported by Burt, reveal important instances of exclusion of pertinent studies as well as ambiguities about what studies were in fact included. In particular, I strongly suspect that Erlenmeyer-Kimling and Jarvik, and thus Jensen, have compiled their data so as to systematically overestimate the IQ correlation of siblings raised together (r_{SIBT}), thus underestimating the true population difference between r_{SIBT} and r_{DZT}, the IQ correlation for fraternal twins raised together. This, in turn, underestimates environmental effects, since the difference ($r_{DZT} - r_{SIBT}$) can result only from nongenetic sources. I reviewed the current status of the scientific scandal surrounding Cyril Burt, noting evidence pertaining to his suspiciously adjusted IQ scores for his presumably separated MZ twin pairs (including questions as to whether many of these pairs even existed); his invariant correlations for both separated MZs and other kinships; the mysterious destruction of his data files; and Dorfman's revelations regarding the Burt (1961) marginal totals and his results on father-child IQ regression to the mean.

I also looked in detail at the Jencks et al. (1972) compilation of kinship data and made the following observations: (1) Jencks et al. appear to overestimate r_{SIBT} by excluding studies that show a low r_{SIBT}. (2) Jencks et al. sometimes confused the different types of IQ tests that Burt presumably used. (3) Jencks et al. rely upon the Erlenmeyer-Kimling and Jarvik compilation, thus perpetuating errors and biases begun there. (4) I uncovered in their compilation arithmetic errors and miscopying of data from original studies. (5) I noted some ambiguity in the criteria that Jencks et al. use for excluding certain studies.

Using the Jencks et al. analysis, I took a close look at results that arise from the use of Jensen's kinship equation. When heritability estimates according to Jensen's equation are compared, what otherwise appears as unordered disarray among estimated h^2 values can indeed be systematically ranked by taking into account differences among kinships in environmental similarity. The discrepancies among h^2 values in Table 2.5 appear to Jencks et al. as merely "dramatic variations in h^2 from one kind of comparison to another" (1972, p. 295). It is partly for this reason that Jencks et al. questioned the usefulness of kinship correlation methods in general, and Jensen's equation in particular. I noted (in Table 2.7) that the inclusion of several studies excluded by Jencks et al. produces a considerably wider range in h^2 values from Jensen's equation than even Jencks et al. suspected— including one value close to zero. According to Jensen's reasoning, equation (2.9), and therefore equations (2.10) through (2.14), should yield roughly equal (certainly not systematically rankable) heritability estimates. Instead, certain of them yield estimates inflated by intrapair environmental similarity. Two kinship categories emerged as important in this connection: the IQ correlation of fraternal twins raised together (r_{DZT}) and the IQ correlation of ordinary siblings raised together (r_{SIBT}). Jensen's equation can also yield absurd h^2 values greater than 1.00 or less than zero. The persistent tendency in the IQ heritability literature to find a sibling IQ correlation of around .50 seems motivated more by attempts to appear scientifically precise than by available data: Even slight deviations in this sibling correlation from a value of .50 can result in absurd h^2 values when using any kinship equation in which the quantity r_{SIBT} appears. Contrary to Jensen, different methods of estimating h^2 do not cross-validate each other. Evidence exists that the crucial assumption of equal intrapair environmental similarity among compared kinships is implausible. In particular, by comparing kinship correlations of equal genetic similarity (in addition to the DZT-SIBT comparisons), it can be seen that non-genetic differences in IQ correlation still exist (as, in same-sex DZTs compared to opposite-sex DZTs; male DZTs compared to female DZTs; male MZTs compared to female MZTs).

Even if h^2 for white and black populations are estimable, no inferences whatever can be made about the heritability of between-group average IQ differences. Few estimates of h^2 for blacks are presently available, and these are inconclusive. Furthermore, no evidence exists for any strong relationship between racial admixture and IQ score.

Two kinds of gene-environment covariance, not distinguished in the literature, may well inflate h^2 estimates via Jensen's equation. Gene-environment interaction may well have this dual character also. (Numerous other assumptions are necessary to estimate h^2; these are considered in detail in the following chapters.) Lacking a clear, consistent, and trustworthy body of pertinent data, and having reason seriously to question certain crucial assumptions in the methodology applied to these data, one must ask: First, is the genetic heritability of human intelligence an estimable quantity; and second, need one postulate the existence of any genes "for" intelligence to account for the observed IQ similarity for biologically related pairs of individuals.

Chapter 3

The Myth of the
Separated Identical Twins

Among the present-day procedures used to estimate IQ heritability, the one with the greatest conceptual appeal uses the correlation between the IQs of identical twins who have theoretically been separated at birth and raised in completely separate and unrelated environments. The argument goes that if in fact the environments of MZ twins are dissimilar, then any similarity between the IQs of the twin pairs is attributable to their identical genes. This IQ correlation of MZ twins raised apart, the statistic r_{MZA} introduced in Chapter 2, has been regarded by most hereditarians as a direct estimate of IQ heritability (h^2). It is thought of by many as being conceptually "purer" than the kinship methods of the sort reviewed in Chapter 2. Indeed, as some have noted, the study of separated identical twins is often regarded as "the single most direct and reliable [way to] estimate . . . IQ heritability" (Layzer, 1975, p. 127). Jensen continues to regard the separated-twin correlation as a crucial technique for estimating IQ heritability (e.g., 1975, p. 177; 1978b, p. 17). The study of separated MZ twins is thus the keystone in the hereditarian's complex edifice.

As it happens, though, twins thought to have been separated early in life and raised in different and dissimilar environments were in truth not so raised. A detailed reanalysis of the original data from three of the four known studies of separated monozygotic twins, which extends Kamin's (1974) seminal analysis, reveals that about two-thirds of the twins were raised in highly similar environments.[1] The IQ correlation for these twins is quite high. But for the remaining twins, those who were in fact raised in reasonably different environments, the IQ correlation is quite low. This means that MZ twins raised in similar environments have similar IQs, and MZ twins raised in different environments have different IQs, despite their identical genes.

The four known separated twin studies, introduced earlier, are those of Burt (1966), involving a sample of 53 English pairs; Shields (1962), involving 37 English pairs; Newman et al. (1937), involving 19 American pairs; and Juel-Nielsen (1965), involving 12 Danish pairs. Burt claimed the IQ correlation of his separated pairs was .86, Shields's correlation was .77, Newman et al.'s was .67, and Juel-Nielsen's was .62, making a grand total of 121 known pairs of separated MZ twins[2]—a small but valuable bit of data.[3] Ever since the first of these studies, countless articles and textbooks in the fields of behavioral genetics, psychology, education, and even sociology, have sustained the myth that these twins were separated virtually at birth and raised apart in ignorance of each other, and yet revealed highly similar IQ scores. Jensen (1969, 1970b, 1972b, 1973, 1975, others) has consistently argued that the IQ correlation of separated MZ twins, based on all four stud- ies, is approximately .80, a false impression vigorously perpetuated by Herrnstein (1971, 1973), Eysenck (1971), and others, such as Loehlin et al. (1975, p. 287 and passim), and Fulker (1975, p. 519). Rao et al. use a correlation of .69 (1976) and Rao and Morton use .68 (1978), based on Newman et al. as their value for r_{MZA}. Sociologists who have delved into the world of IQ heritability estimation, such as Eckland (1967, pp. 177–78; 1973, pp. 86–87) and Jencks et al. (1972, pp. 309–16, who use the Newman et al. study) continue to treat these twins as though they were raised in separate environments.

Of the four studies in question, only Shields, Newman et al., and Juel-Nielsen can be examined in detail here. Each of these studies contains detailed appendixes and case descriptions, evidently largely ignored over the years, which to some extent permit a systematic analysis of the extent of environmental similarity characterizing the separated MZ twin pairs. The studies indicate whether or not the twins attended the same school after they were separated; whether or not the twins were reunited after their initial separation; some infor- mation on whether their respective adoptive homes after separation were similar in terms of occupational, educational, income, or cultural advantages; and so on. Burt's study (Burt, 1955, Burt and Howard, 1956; Burt, 1958, 1966, 1971, 1972) contains no case descriptions and obviously cannot be used in this analysis—even apart from the suspi- cions already raised about the quality of Burt's data.

To estimate IQ heritability from the study of separated MZ twins, it is crucial that the researcher make sure that environmental effects are held constant to the extent permitted by the data (the ceteris paribus requirement). This means that in the study of separated identical twins, who have exactly the same genes, any similarity between their IQs may indeed be attributed to their genes if: (1) the twins were separated at birth; (2) they were raised after separation but prior to testing in completely different, unrelated, families in uncorrelated social environments; and (3) after separation, members of all pairs were spread over a wide range of different environments.[4] This chapter is largely concerned with the violation of this one requirement.

I explore four ways in which inadvertent environmental similarity between the presumably separated identical twins came about, thus producing artificial inflation of their IQ correlation (r_{MZA}) as an estimate of IQ heritability. These four sources of environmental similarity, unanalyzed by Shields, Newman et al., and Juel-Nielsen, are[5]

1. *Late separation.* The later the age of separation of the twin pairs, the longer they are exposed to a common familial, educational, and sociocultural environment. One might thus compare the IQ correlation for twins separated very soon after birth to the IQ correlation for twins separated relatively later.

2. *Reunion prior to testing.* Certain of the twins in these studies were indeed separated early (within their first year) but were reunited prior to being tested for IQ. Twins who are so reunited are exposed to a common environment for a longer period than twins who are not reunited.

3. *Relatedness of adoptive families.* While most of the twins were indeed placed in different families after their initial separation, in the vast majority of cases the twins were placed in different branches of the same family, or in families of close friends, or in the same orphanages. A common pattern was for one twin to be raised by its natural mother and the other by some member of the extended family, such as an aunt, uncle, or grandparent. Such twins share more environmental similarities than do twins who are raised in completely unrelated families.

4. *Similarity in social environment after separation.* Twins can indeed be separated on the day of their birth, be raised in unrelated families, and not be reunited prior to testing, but nevertheless be raised in environments that are similar in educational or socioeconomic characteristics, or in environments that permit frequent personal contact between the two. For example, both twins could attend the same school, or attend schools of the

same type or quality for the same number of years in the same locality, or have foster parents of similar (or even identical) occupations, educations, or incomes, or see each other every day. Later I compare the IQ correlations of such twin pairs to those who had fewer similarities in their social environments.

THE SAMPLES AND SAMPLE BIAS

The three studies of separated twins being considered constitute survey research, and as such, they represent attempts to generalize from some sample of cases to some broader hypothetical population from which the cases have presumably been drawn. The hypothetical population in question would be defined technically as all pairs of monozygotic twins presumed to exist, in a given culture or country, and presumed to be separately raised, when the study in question was done. A true random sample of pairs from such a population would result in a reasonably high probability of the sample being representative of the population, especially for larger samples. In all of the separated twin studies, however, the samples were anything but random, and the original authors make no claims of having sampled randomly. The samples consisted mainly of twins who responded to radio and newspaper appeals. Since monozygotic twins who claim to have been raised separately are obviously rare, the samples are small. One thus needs to be concerned with the issue of whether the samples can be expected to be reasonable approximations of their hypothetical populations.

It should be remembered that sample characteristics and population characteristics are, at least conceptually, distinct. Whether or not the separated twins actually sampled were themselves characterized by a high or low degree of environmental similarity is one issue (which I take up in the next section). Whether or not the investigators' sampling procedures inadvertently biased the sample in a direction of greater (or lesser) environmental similarity in comparison to a broader population of separated MZ twins is another. I turn now to a brief look at the data-gathering procedures of the three studies.

The Shields (1962) study, largest of the three, was done in England between 1953 and 1962 and involved an initial sample of 44 separated MZ pairs in addition to 34 nonseparated MZ pairs (MZTs). Shields obtained an IQ correlation of .77 for 37 separated pairs, and a virtually identical correlation of .76 for the nonseparated pairs, a value quite a bit lower than that usually found for MZTs. (One plausible reason for the closeness of these two correlations was that the majority of Shields's separated pairs were never in any real sense separated at all.) Only those 37 pairs Shields designated as "raised separately" are discussed here. (Seven of the initially sampled 44 separated pairs were not used for various reasons, such as having obtained an IQ score for only one member of a pair.)

There are 13 male pairs and 24 female pairs in the sample, ranging in age from 8 to 59 years, with a median age of 40 years. Shields located these twins by means of a BBC radio broadcast in 1953, with a special appeal "for identical twins brought up apart to come forward in the interests of scientific research" (Shields, 1962, p. 21). Consequently, Shields had to rely largely upon the twins' own accounts of their separation, and exclusively upon a sample of volunteers responding to the initial radio appeal. This raises at least two serious problems. First, it is entirely possible that some twins might have exaggerated the extent of their separation for purposes of being in the study, which would bias the sample in the direction of overrepresentation of environmentally similar pairs. Second, the use of volunteers introduces a potential for another kind of error: The twins had to know that they were in fact twins to respond to Shields's radio broadcast. Twins raised separately who did not even know of their twinship would obviously not have responded to the radio appeal and were therefore excluded from the sample. To the extent that such excluded twins might have had greatly differing environments, the sample may well be biased toward underrepresentation of environmentally dissimilar pairs. Both of these elements undoubtedly resulted in some sampling bias, but unfortunately, the amount of bias is not ascertainable.

Shields established the zygosity of the pairs after a screening interview. His zygosity diagnosis appears to be quite reliable, as the twins were considered to be monozygotic only if they were concordant on the following five presumably highly heritable criteria: blood grouping (in both ABO grouping and Rh factor, as well as in other systems such as the Kell, Lewis, and Duffy); ability to taste PTC; color blindness; fingerprint pattern; and anthroposcopy (visual inspection of such characteristics as body build, eye color, facial features, and so on). True dizygosity can be misdiagnosed as monozygosity and vice versa, but there is no reason to suspect Shields of either error.

The IQ test given by Shields is not usually regarded as an IQ test. It consisted of two tests: One, called the Dominoes test, is a nonverbal test developed by British psychologists during World War II, which requires the testtaker to fill in the number of dots on a blank domino that would complete a logical series with other dominoes given for an item. The test contains 48 items, with 1 point scored for each correct item. According to Shields a raw score of 28 points corresponds to an IQ of 100 (1962, p. 59). The other test was part of Raven's Mill Hill Vocabulary Scale (Synonyms Section, Set A), a test consisting of synonym identifications. Shields estimated that a raw score of 19 points on the Mill Hill corresponded roughly to an IQ of 100. He found the standard deviation for the Dominoes to be about twice that of the Mill Hill, for both his separated and his nonseparated twin samples. As a correction for this, his final raw IQ score consisted of the raw score on the Dominoes plus twice the raw score on the Mill Hill Vocabulary (1962, p. 60). These composite scores will be used in the analysis later.

Shields's correlation of .77 for the separated MZ sample was based upon these composite raw scores. He made no attempt to transform them to a mean of 100 and a standard deviation of 15, as is the common practice in the world of IQ testing,[6] which seems wise, given the lack of appropriate age and sex standardization for both the Dominoes and the Mill Hill (Kamin, 1974, pp. 47–48). This, however, did not prevent Jensen from transforming Shields's raw scores to a mean of 100 and a standard deviation of 15 separately for the two tests, with the result that the standard deviation for the composite scores was 13.4 points (1970b, p. 135). The standard deviation for Shields's untransformed composite scores is 18.4. The lower standard deviation arising from Jensen's transformation means that transforming the scores decreases the overall variability, thus increasing the overall intrapair IQ similarity, of the twin pairs. This *can* inflate the double-entry or intraclass IQ correlation (r_{MZA}), but it will definitely inflate the so-called difference correlation, r_d, advocated and used in 1970 by Jensen, which is as follows:

$$r_d = 1 - (|\bar{d}_k| \; / \; |\bar{d}_p|)^2, \qquad (3.1)$$

where

$|\bar{d}_k|$ = mean of the absolute intrapair differences; and
$|\bar{d}_p|$ = mean absolute difference between all possible paired comparisons in the general population, estimated as 1.13σ.[7]

This difference correlation can be estimated if one has some estimate of σ, the standard deviation for the general population in question. Assuming σ to be 15 IQ points yields an estimate of $1.13 \times 15 =$ 16.85 or about 17 IQ points, for $\bar{d}_p$. One can interpret this 17-point IQ value as the extent to which one would expect ordinary unrelated pairs of people to differ from each other on the average in their IQs. Reducing the overall standard deviation of raw scores in a sample also necessarily reduces the value of $\bar{d}_k$, but it will not reduce $\bar{d}_p$, thus artificially inflating the correlation r_d. The intraclass (and double-entry) correlation for Shields's sample is .77. But the r_d as calculated by Jensen using his transformed scores is .84 (1970b, p. 137). The use of r_d on the Newman et al. and Juel-Nielsen samples is even more inflationary: The intraclass correlation for Newman et al. is .67; Jensen's r_d is .76. For Juel-Nielsen, the intraclass correlation is .68, and Jensen's r_d is .86. (Incidentally, this flawed statistic, r_d, was used in Munsinger's 1977 study of birth-weight and IQ differences for twin pairs. Jensen's difference correlation was calculated for several samples of MZ twins raised together, but, as noted by Kamin, 1978, p. 166, he used signed differences in the numerator rather than *unsigned*, absolute differences, thus adding arithmetic error to methodological flaw.)

The Newman et al. (1937) study of 19 separated MZ pairs, the oldest of the three, is the only study of separated identical twins ever done in the United States. Completed over a ten-year period in Chicago, the study also contained 50 nonseparated MZ pairs and 50 nonseparated DZ pairs. The investigators obtained an IQ correlation of .67 for the separated twin sample (erroneously reported by Jensen, 1969, p. 52, as .77). (The correlations for the nonseparated MZ and DZ samples were .89 and .63 respectively; see Table 2.3.) The separated MZ sample contained 7 male pairs and 12 female pairs, ranging in age from 11 to 59 years, with a median age of 26 years. The popular 1916 version of the Stanford-Binet IQ test was used.

In a manner similar to that of Shields, Newman et al. attempted to locate by radio and newspaper advertisements identical twins who had been raised separately. They requested responses from any twins who thought they qualified as being separated and "who were so strikingly similar that even your friends or relatives have confused you" (Newman et al., 1937, p. 135). (One is tempted to ask how separated such twins really were if "friends or relatives have confused" them.) After this initial screening, potentially monozygotic pairs were invited to Chicago and subjected to tests for zygosity, which consisted of observations of eye color, stature, direction of hair whorl, and other such traits, and an examination of palm prints and fingerprints. The Newman et al. study could not use either blood-grouping diagnosis or ability to taste PTC since at that time neither of these methods for zygosity diagnosis was known.

Their method of diagnosing zygosity therefore introduced a poten-

tial source of sample bias not present with the Shields study. It is entirely possible, though not very likely, for true MZ twins to differ in
such characteristics as height, stature, facial features, and other traits
seen upon visual examination. This means that any physically dissimilar MZ twins would have been eliminated from the final sample. In
this sense the sample is biased. To the extent that true monozygotic
twins who differ physically are also treated differently by people in
the environment, the Newman et al. sample underrepresents environmentally dissimilar monozygotic twins.

In their attempt to reach out as far as possible for monozygotic
twins who by their own account were raised separately, Newman et
al. introduced yet another unfortunate bias into their sampling procedure. They promised all potentially separated MZ pairs a free trip to
the Century of Progress Exposition in Chicago: "Pair after pair, who
had previously been unmoved by appeals to the effect that they owed
it to science . . . to permit us to study them, could not resist the offer
of a free, all-expenses paid trip to the Chicago Fair" (1937, p. 134).
Clearly, such an attractive offer, in the midst of the Great Depression,
might well have caused any twins appearing in the final sample to exaggerate the extent to which they were separately raised, which could
have resulted in overrepresentation of environmentally similar twins
in the final sample.

A third problem exists with the Newman et al. sample: Because of
the cost, twins who were separated from each other by great geographic distances were systematically excluded from the sample, yet
such twins might have been culturally or environmentally separated
as well. Exclusion of such pairs could well bias the sample toward
overrepresenting environmentally similar twins. One potentially MZ
pair was eliminated because one member of the pair was "now in
Denmark and did not speak English" (1937, p. 134). In another instance, "one twin lived in Alaska and the sister in California" (1937, p.
134). In still another, "the twins were separated by the whole width of
the American Continent" (1937, p. 134).

Needless to say, the Newman et al. method of radio and newspaper
recruitment brought some humorous responses. One man, a stock-
broker from Canada, responded to the initial appeal by indicating that
he and his twin brother were just what was wanted. He also mentioned that he had some mining stocks selling at a "ridiculously low
figure." He urged Newman, Freeman, and Holzinger to avail themselves of this unusual opportunity. The investigators expressed no in-

terest in the stocks. The twins then refused to participate in the study, even after Newman et al. expressed a mild interest in the stocks.

Another pair, two "maiden-lady twins in South Carolina," wrote that they would gladly offer themselves at the altar of science "for any sort of experiment" if only the investigators would support them for "the years during which they were made use of" (1937, p. 133). A third offer, reminiscent of Watsonian behaviorism, came from a charitable organization which was willing to present Newman et al. with a pair of orphaned twin infants, who they "could then separate and bring up in environments as different as we might wish" (1937, p. 133).[8]

In sum, the Newman et al. sample was subject to at least three sources of bias arising from the sampling procedures, any one of which could inflate any IQ correlation based on the whole, or any part of, the sample: (1) An indeterminate number of physically dissimilar but true monozygotic pairs were eliminated. (2) The inducement of a free trip to the Chicago Fair, in the midst of the Depression, might well have caused certain twins to exaggerate the degree to which they were separately raised. (3) Twins who were potentially monozygotic but separated by great geographic distances, thus perhaps by cultural or environmental distances as well, were excluded. Finally, it should be noted that as with Shields, any twins appearing in the final sample had at least to be aware of each other's existence.

Juel-Nielsen's (1965) 12 Danish pairs contained 9 male and 3 female pairs ranging in age from 22 to 77, with a median age of 50 years. The Juel-Nielsen study is the most recent and contains lengthy case descriptions for each twin pair. Nine of the 12 pairs were tested twice for IQ using a Danish translation of the Wechsler Adult Intelligence Scale (WAIS). The test-retest correlation reported by Juel-Nielsen (1965, p. 106) was .89, slightly less than that ordinarily obtained for the Wechsler scales. (As Juel-Nielsen, p. 54, indicates, the WAIS had never been standardized on any Danish population. As Kamin notes, 1974, pp. 52–65, failure to standardize for age is a problem with all of the separated twin studies. This issue is discussed later.)

Jensen's (1970) analysis combined the $time_1$ and $time_2$ scores for each individual, such that IQ score (for a single person) = $(IQ_1 + IQ_2)/2$. This will slightly inflate the overall correlation for the sample, since combining the scores reduces the overall intrapair differences. For the sake of comparability to the other studies, it would make more sense to base the IQ correlation on only the first administration, as Juel-Nielsen himself did, since the other separated twin studies involved only one test per individual. Jensen, however, reported a correlation, based on combined time$_1$ and time$_2$ scores, of .68 (1970b, p. 137). The correlation for time$_1$ scores alone is .62. It is somewhat lower than that for time$_2$ scores alone, which is .68. I analyze time$_1$ and combined scores separately.

Between 1955 and 1959, Juel-Nielsen obtained 8 of his twin pairs from a registry of twin births maintained by the Institute of Human Genetics at the University of Copenhagen. The 4 remaining pairs were obtained from other sources, one of which was a journalist who became aware of the pair as a result of their reunion after many years of separation. Zygosity was established by means of blood groupings, eye color, hair color, fingerprints, and ability to taste PTC. Juel-Nielsen's zygosity diagnosis was thus about as reliable as Shields's.

Juel-Nielsen's sampling procedures were thus superior to those of Shields and Newman et al. The sample did not consist of volunteers, and the zygosity diagnoses were not likely to eliminate physically dissimilar monozygotes. Furthermore, the Juel-Nielsen twins, few though they were, appear to have been somewhat more completely separated than was the case in the other two studies, and the Juel-Nielsen correlation (time$_1$ scores) of .62 is less than Shields's .77 and Newman et al.'s .67. Still, as Juel-Nielsen himself indicates, "the geographical distance between childhood homes was on the whole small . . . great diversities in environment are not to be expected" (1965, p. 97). He notes further that even for the 5 pairs of twins revealing the greatest relative socioeconomic differences between foster homes, "these differences were by no means extreme" (p. 102). Juel-Nielsen, Shields, and Newman et al., consistently caution the reader against inferring too much environmental separation for the twin pairs in their samples, but their cautions tend to be just as consistently ignored by Jensen and most other contributors to the professional literature on IQ heritability.

A CLOSE LOOK AT THE DATA

Tables 3.1 through 3.4 present the IQ correlations for the Shields, Newman et al., and Juel-Nielsen twins while controlling for the four environmental variables for which there is adequate evidence in the descriptions of the original studies: age at separation; whether or not a childhood reunion occurred prior to testing; whether or not the twins were raised in related family environments; and extent of social environmental similarity for members of a twin pair. How these varia-

Table 3.1. **IQ Correlations and Mean Intrapair Differences, by Age of Separation**

	Late (> 6 months)			Early ($\leq$ 6 months)		
	r	Difference	N	r	Difference	N
Shields						
Raw scores	.71	10.07	14	.79	9.09	23
Transformed						
scores	.71	8.29	14	.79	7.30	23
Newman et al.	.71	10.18	11	.48	5.50	8
Juel-Nielsen						
Time$_1$ scores	.58	7.75	8	.04	6.25	4
Combined						
scores	.70	6.88	8	.15	5.50	4
Averaged						
correlation[a]	.71		33	.61		35

[a] Calculated as a weighted average, $\Sigma rn/N$, using transformed scores for Shields and combined scores for Juel-Nielsen.

bles were measured will be described in turn. (Case-by-case codes for each variable are given in Appendix B.)

Separate IQ correlations are calculated within the categories of each variable. (For example, in Table 3.1 separate correlations are given for "late" and "early" separation.) The higher the correlation, the more similar the twin pairs are in their IQs. Correlations are calculated separately for Shields's raw scores, Shields's transformed scores (see Chapter 3, note 6), Newman et al. scores, and Juel-Nielsen's time$_1$ and combined [(time$_1$ + time$_2$)/2] scores. Also included are the mean absolute intrapair IQ differences for each category. The smaller this value, the more similar the twin pairs are in IQ. Within a category of a control variable, and across the three studies, averaged correlations are shown.[9]

Late Separation

The longer the period the twins spent together in the same home before their separation, the greater the extent to which they shared a common environment in infancy or early childhood. Under an environmental hypothesis, this would mean that twins separated at a relatively older age would reveal a higher IQ correlation than twins separated at a relatively younger age.

The data reveal little overall effect of age at separation upon the IQ correlation (Table 3.1). Various analyses were done using different cutting points to define early and late separation: at birth versus any time after birth; at 6 months or less versus more than 6 months; at 12 months or less versus more than 12 months; at 18 months versus more than 18 months; and finally, at 24 months versus more than 24 months. No clear effects on either the IQ correlation or intrapair differences were seen. Results are given in Table 3.1 using the median age at separation as determined across all three studies (6 months). For each study, the IQ correlation for twins separated relatively early (at 6 months or less) does not differ significantly from the IQ correlation for those separated later. The averaged correlations do not differ significantly.

If there is any effect at all, it is in the direction of an environmental hypothesis: The IQ correlations for the late separations are generally higher, a conclusion contrary to that reached by Vandenberg (1971, pp. 191–92; Vandenberg and Johnson, 1968). Vandenberg analyzed these same data and concluded that the later the separation of the twins, the less their IQ similarity. There are at least five reasons why our results differ from Vandenberg's, the first three of which evidently result from errors on Vandenberg's part. First, for reasons that are not at all clear, Vandenberg did not include any of Shields's pairs in his analysis, even though the Shields study does appear in his references (Vandenberg, 1971, p. 217). The Shields pairs are analyzed, however, in Vandenberg and Johnson (1968). These pairs show slightly less IQ similarity for the late separated pairs. Second, he includes only 18 of the Newman et al. 19 pairs—a transcription error. Third, he shows Juel-Nielsen with 13 pairs, having counted one Newman et al. pair as a Juel-Nielsen pair (confirmed by Vandenberg, personal communication, 1980). Fourth, he included 7 pairs from isolated case studies of separated twins.[10] Fifth, his analysis is based on absolute intrapair differences rather than correlations. Despite the errors, the Vandenberg finding continues to be reported in textbooks on behavioral genetics (e.g., McClearn and DeFries, 1973, p. 286).

Although it does not appear that lateness significantly affects the IQ correlation of separated MZ twins, there is a tendency in the Newman et al. and Juel-Nielsen studies for the late-separated pairs to be more similar (in terms of the correlation) in IQ. Jensen and the others give their readers the impression that the twins were separated either at birth or very early in life, thus leading one to believe that their IQ correlation, and thus the IQ heritability estimate, is based on such twins. But as Table 3.1 clearly shows, 33 out of the 68 pairs—almost half—were separated after 6 months. More important, among these, 20 were separated at or after 18 months.

In his article on these twins, Jensen reported that "all of Shields' twins were separated before 6 months of age" (1970b, p. 135), a gross error of fact as is easily seen from Table 3.1. Indeed, as Shields's descriptive appendix clearly indicates, several pairs were actually separated quite late: One pair, called Adeline and Gwendolyn, were separated at $9\frac{1}{2}$ years (1962, p. 244). Another pair, Molly and Dorothy, were separated at age 8 (p. 206); and still another pair, Pauline and Sally, at age 7 (p. 208).

Of the Newman et al. study, Jensen has said: "In 9 [cases the age of separation] was less than six months" (1970b, p. 136). That is incorrect. Newman et al. (1937, p. 144) clearly indicate that only 5 pairs were definitely separated before six months of age. Newman et al. are uncertain of the exact age of separation of 3 cases, listing it as "about 6 months" (p. 144). But even the inclusion of these questionable pairs gives 8 who were separated at or before 6 months, not 9. As Table 3.1 shows, the remaining 11 pairs were separated after 6 months of age with 1 pair (Augusta and Helen) not separated until after their sixth birthday.

Two things now seem clear: (1) Regardless of how one defines lateness of separation, it does not appear to affect the IQ correlation significantly, although a slight tendency for the later separated pairs to be more similar can be noted. (2) Across all three studies, almost half (33 of the 68 pairs) were separated after 6 months, with most of these (20) being separated at or after 18 months. Consequently, the environments of these twins cannot, at least by this criterion, be considered uncorrelated.

Reunion Prior to Testing

Although virtually all of the 68 twin pairs across the three studies were indeed separated in some manner for at least some period, most of them—44 or nearly two-thirds—were reunited for some period before being tested for IQ. After their initial separation from either one or both of their natural parents, they were brought back together, either into a single family under the same roof, or by means of frequent visits to their respective adoptive families, and then separated once again. Jensen, Herrnstein, and Eysenck, and indeed their critics, make no note of this. Reunion prior to testing is clearly a potential source of environmental similarity, to which any similarity in IQ can at least in part, be attributed.

For purposes of analysis, a twin pair was classified as having been reunited if at some point after the initial separation from the natural parents, but before reaching the age of 21 (or before being tested for IQ, whichever was sooner), the twins were again brought back together in the same home, or if it was clear that they met on a fairly regular basis, such as once a week, or once a month, or for summer vacations, holidays, and the like.[11] It is not possible to take into account the length of the reunion, for this is not systematically reported in any of the three studies. I distinguish here only those twins for whom there is some evidence of a reunion of whatever sort from those for whom there is no such evidence.

Shields's detailed appendix tells a story ignored by Jensen: Richard and Kenneth "lived only a few miles away" from each other, and "from the age of 9 met once a week" (Shields, 1962, p. 163). They were tested at age 14. Bertram and Christopher at the time of the study were "living next door to one another in the same Midlands colliery village" and were "constantly in and out of each other's houses" (p. 164, italics added), yet Bertram and Christopher are considered to have been raised separately in the literature on IQ heritability.

Rodney was separated from his mother and his twin brother Barry at birth. Barry stayed with the natural mother. Rodney stayed instead with a paternal aunt "until the age of 9 when . . . he was returned to mother against his will" (p. 175). Herbert and Bryan lived apart but "met for their summer holidays, knowing they were twins" (p. 187); Christine and Nina "generally spent summer holidays together in each other's homes" (p. 219); Madge was "taken every week to visit her twin," Olive (p. 198); Joan and Dinah "were reunited most of the time from 5 to 15" (p. 214, italics added); and so on for 27 out of Shields's 37 twin pairs.

The same is the case for 11 of Newman et al.'s 19 pairs—Ada and Ida saw "a great deal of each other" (p. 203); Harold and Holden "have seen a good deal of each other" (p. 144); Maxine and Virginia had "a few visits" (p. 289)—and for 6 of Juel-Nielsen's 12 pairs. That gives a total of 44 out of the 68 pairs who are considered by much of the literature to be raised in "uncorrelated environments."

Table 3.2 clearly shows the effect of reunion versus nonreunion on the IQ correlation: Twins who were reunited show considerably more IQ similarity than those who were not. For the Shields study, the cor-

Table 3.2. **IQ Correlations and Mean Intrapair Differences, by Childhood Reunion**

	Reunited			Not reunited		
	r	Difference	N	r	Difference	N
Shields						
Raw scores	.82	6.74	27	.52[a]	16.80	10
Transformed						
scores	.86	5.44	27	.55[a]	13.70	10
Newman et al.	.87	4.55	11	.51[a]	13.25	8
Juel-Nielsen						
Time₁ scores	.67	7.16	6	.51	7.33	6
Combined						
scores	.77	5.00	6	.59	7.83	6
Averaged						
correlation	.85		44	.55[b]		24

[a] $P < .10$, compared to the correlation coefficient to the left, using z-transformations.

[b] $P < .01$.

relations of the raw scores for the reunited pairs are .82, while those for the twins who were not reunited are .52. The difference between these two correlations is statistically significant at a borderline level ($P < .10$), despite the smallness of the samples. The picture is much the same when Shields's raw scores are transformed to yield an overall mean "IQ" of 100 and a standard deviation of 15 (see the second row of Table 3.2): Reunited twin pairs differ considerably less from each other (by an average of 5.44 IQ points) than do those not reunited (13.70 IQ points). Twins not reunited differ from each other by an average of about 14 IQ points, fairly close to the 17 points by which pairs of ordinary unrelated persons selected randomly from a population are expected to differ from each other. Using these transformed scores, the IQ correlation for reunited twins is .86; for those not reunited, .55.

The story is strikingly similar for Newman et al.'s 19 twin pairs. Those who were reunited (11 pairs) differed by an average of only 4.55 Stanford-Binet IQ points, while those not reunited differed by an average of 13.25 points. The IQ correlation for the reunited twins is .87; for those not reunited it is .51. The effect of the "intelligence genes" of MZ twins upon their IQs seems to depend heavily upon whether or not they see each other periodically. (The same tendency exists with the Juel-Nielsen twins, although differences are not statistically significant.)

Finally, averaging across all three studies yields a correlation for the 44 reunited pairs of .85; for the remaining 24 unreunited pairs, it is .55, a statistically significant difference ($P < .01$). Thus, the similarity in the IQs of pairs of separated identical twins is seen to depend upon whether or not they are reunited after their initial separation, evidence of a strong environmental effect on IQ. Using the more separated among the separated MZ twins to estimate IQ heritability would reduce the Jensen-like h^2 estimate from .80 to .55.

Relatedness of Adoptive Families

From the descriptive data provided by Shields, Newman et al., and Juel-Nielsen, it appears that presumably separated MZ twins were raised in four kinds of correlated family environments: (1) Some twin pairs were raised in different branches of the same family, as when one twin was raised by a natural parent, the other by some member of the extended family, such as an aunt, uncle, or grandparent; in some cases, both twins were raised by members of the same extended family. (2) In some cases, one twin was raised by a natural parent, the other by friends of the family, or both twins were raised by friends of the family. (3) In some cases, one twin was raised by some member of the extended family, the other by friends of the family. (4) In a few cases, both twins spent some time in the same orphanage prior to adoption into separate adoptive families. By far the most common pattern, especially in the Shields study, was for one twin to be raised by a natural parent (the mother), the other by some member of the extended family, usually an aunt or the maternal grandmother.

Twenty-nine of Shields's 37 twins pairs, 11 of Newman et al.'s 19 pairs, and 8 of Juel-Nielsen's 12 pairs, were raised in one of the above four ways (see Table 3.3). Thus 48 pairs, more than two-thirds of the total, were raised in family environments that were clearly related in some manner. The remaining 20 pairs were either raised in two completely different and unrelated families, or one twin was raised by a natural parent (or extended family member) while the other was raised in a separate family unrelated and unknown to the natural family. While the Shields appendix is quite precise as to which twin was raised by what kind of family, the Newman et al. and Juel-Nielsen appendixes are often vague, frequently indicating only that one or both

Table 3.3. **IQ Correlations and Mean Intrapair Differences, by Relatedness of Adoptive Families**

	Definitely related			Possibly related		
	r	Difference	N	r	Difference	N
Shields						
Raw scores	.83	7.79	29	.45[a]	15.50	8
Transformed						
scores	.83	6.28	29	.44[a]	12.75	8
Newman et al.	.66	7.90	11	.76	8.62	8
Juel-Nielsen						
Time$_1$ scores	.56	7.75	8	.59	6.25	4
Combined						
scores	.70	5.88	8	.64	7.50	4
Averaged						
correlation	.77		48	.61		20

[a] $P < .10$, compared to the correlation coefficient to the left, using z-transformations.

twins were adopted, but not by whom. Consequently, the families of these 20 pairs are designated in Table 3.3 as "possibly related."

Although Jensen has said that "*some* separated twins are reared . . . in different branches of the same family" (1970b, p. 134, italics added), the Shields appendix clearly shows that exactly 23 pairs of the 37 were raised in different branches, with one twin raised by a natural parent, the other by some member of the extended family. This raises the question of what constitutes "some" and what constitutes "most."

Despite difficulties with the descriptive data, it can still be seen that twins reared in families that were only "possibly related" reveal less IQ similarity than those who were reared in definitely related families. The 29 pairs of Shields's 37 who were definitely raised in related families differed on the average in their IQs by about 6 transformed IQ points (6.28). Their IQ correlation is .83, for both raw and transformed scores. For the 8 pairs raised in possibly related families, the average IQ difference was nearly 13 points for transformed scores—not too far from the 17-point difference expected for pairs of totally unrelated individuals in a general population. The IQ correlation for these twins is low, .45 for raw and .44 for transformed scores.

While no significant differences between "definite" and "possible" exist for the Newman et al. and Juel-Nielsen studies, the difference between the averaged correlations (.77 for pairs in definitely related families; .61 for pairs in possibly related families) approaches statistical significance. In general, then, relatedness of adoptive families does appear to affect the IQ correlation of separated MZ twins. This effect is most pronounced in the Shields study, where the more accurate descriptive accounts are given. Once again, I surmise that estimates of an IQ correlation of .80 for separated identical twins are inflated.

Similarity in Social Environment

The similarity in educational, socioeconomic, and interpersonal environments, referred to here as *social environment,* is a central reason why monozygotic twins regarded in the professional literature as separately raised reveal similar IQ scores. MZ twin pairs who have had similar social environments (such as similar schooling) have similar IQs, and twin pairs who have relatively different social environments (especially different schooling) have different IQs.

Similarity in social environment between members of a given pair was coded as follows: A pair of twins is classified in Table 3.4 as having "strong" similarity in social environment if both twins attended the same school for any length of time, or if they attended different schools of the same type (public, private, or parochial) for exactly the same number of years. A pair was also classified as "strong" in social environmental similarity if they lived close to one another and if it was clear from the descriptive appendix that they had virtually daily personal contact. The social environment of a pair was considered to be "possibly similar" (perhaps "moderately similar" would have been a better term) if they attended different schools (for a different number of years) but were raised in homes with similar or identical socioeconomic characteristics (if, for example, the adoptive fathers or household heads had similar or identical occupations, incomes, or years of formal education). While the socioeconomic characteristics of the twins' adoptive homes are not reliably or consistently reported in any of the three studies, schooling (for the twins themselves) was quite consistently reported, and therefore schooling was used as the central criterion for deciding between the "strong" and the "possible" categories.

Similarity in social environment was considered "minimal" if the pair had never attended the same school, differed in years of school-

Table 3.4. **IQ Correlations and Mean Intrapair Differences, by Similarity in Social Environment**

	Strong or possible[a]			Minimal		
	r	Difference	N	r	Difference	N
Shields						
Raw scores	.89	6.26	27	.45[b]	17.80	10
Transformed						
scores	.89	5.07	27	.44[b]	14.70	10
Newman et al.	.91	3.92	12	.36[b]	15.57	7
Juel-Nielsen						
Time$_1$ scores	.66	6.71	7	.50	8.00	5
Combined						
scores	.72	5.14	7	.51	8.20	5
Averaged						
correlation	.87		46	.43[c]		22

[a] Separate "strong" and "possible" correlations are: for Shields's raw scores, .90 (14) and .87 (13); for Shields's transformed scores, .90 (14) and .87 (13); and for Newman et al., .71 (4) and .94 (8). Only 2 Juel-Nielsen pairs were classified as strong. The correlation for time$_1$ scores for the 5 "possibles" was .63; for their combined scores .71.

[b] $P < .05$, compared to the correlation coefficient to the left, using z-transformations.

[c] $P < .001$.

ing, did not live in homes with similar socioeconomic characteristics, and in any other way were clearly quite "separate" (if, for example, they were raised in different towns or even countries, with no personal contact). The analysis here focuses mainly on these minimally similar pairs.[12] Case-by-case codes for all pairs appear in Appendix B.[13]

I begin with the Shields study. Bertram and Christopher grew up "amicably, living next door to one another. . . . They went to the same school . . . both twins became garage mechanics" (1962, p. 165). They are virtually identical in their IQs (58 and 64 raw; 95 and 98 transformed). Bertram and Christopher, treated by Jensen and other hereditarians and in textbooks, as being "raised separately," are classified in Table 3.4 as having "strong" environmental similarity.

Other examples of strong similarity in the social environments of the "separated" Shields twins abound in the Shields appendix. A few of these are listed in Table 3.5. Benjamin and Ronald, who used to

Table 3.5. **How Separate Were the Social Environments of the Separated Identical Twins?**

Shields (1962)

Bertram and Christopher: Grew up "amicably, living next door to one another"; "They went to the same school"; "both twins became garage mechanics" (pp. 164–65).

Timothy and Kevin: "Attended the same school"; "lived a few roads away from each other in the same Northern industrial town" (p. 182).

Benjamin and Ronald: "At school together"; "have continued to live in the same village"; "used to truant together" (pp. 188–89).

Jessie and Winifred: "Brought up within a few hundred yards of one another"; "gravitated to one another at school at the age of 5"; "play together quite a lot at school and during the evenings" (pp. 189–91).

Molly and Dorothy: "Continued to attend the same school"; "since marriage they have lived in the same Midlands town and meet regularly" (p. 206).

Joan and Dinah: "Quarrelled a lot as children"; "together most of the time" (pp. 214–15).

Charlotte and Laura: "Lived as close neighbours"; "attending the same school until they were 15"; "were closely attached" (p. 225).

Mary and Nancy: "Attended the same school"; "have lived in the same country village" (p. 227).

Joanna and Isobel: "Went to private schools together"; "both [adoptive] homes were strict . . . and highly religious" (p. 233).

Frederick and Peter: "Different schools" but "in the same Kent town"; "in the window cleaning business together" (p. 172).

Foster and Francis: "Went to different schools" but "brought up about 5 miles apart"; "would meet in each other's homes" (p. 173).

Amy and Teresa: Different schools but "social conditions in both homes were very poor" (p. 237).

Dora and Brenda: Different schools until the age of 12; but "special point was made of seeing that the twins had the same clothes, the same presents and the same pocket-money" (p. 240).

Newman et al. (1937)

Edwin and Fred: "In the same New England town"; "went to the same school for a time"; "families were of essentially the same social and economic status"; "each owns a fox terrier dog named Trixie" (p. 281).

Kenneth and Jerry: Jerry's foster father was "a city fireman with only a fourth grade education"; Kenneth's was "a city fireman with a very limited education"; they "lived in the same city"; "no difference in the schooling of these two boys" (pp. 264–66).

Harold and Holden: "Attended the same township high school and were together in this school for three years"; "educational environments may be considered essentially equal" (p. 228).

Table 3.5 *(continued)*

Ada and Ida: "Have seen a good deal of each other, sometimes living together for months at a time"; "Each had the equivalent of third-grade training" (pp. 203–4).

Maxine and Virginia: "Only minor differences in the social milieus of the two girls" (p. 290).

Thelma and Zelma: Thelma's foster father "had been a schoolteacher"; Zelma's "had taught school for a time" (p. 256).

Edith and Fay: "Were in their youth so similar in every way that one frequently substituted for the other at her place of employment" (p. 139); "no significant difference in the character of the two foster-homes"; "two homes were in small towns and definitely similar in social and cultural status" (p. 195).

Juel-Nielsen (1965)

Karin and Kristine: Different schools but "SES circumstances of foster homes can hardly have differed very much" (p. 229); they "often met at dances and the like" (p. 223); both were domestics; both married "smallholders" (p. 229).

Ingegerd and Monika: Attended the same school (p. 113) where they were often confused with each other (p. 104); and "spent the latter half of their childhood together with their mother" (p. 90).

Martha and Marie: Grew up in the same neighborhood (p. 148); attended the same school for the same number of years (p. 151).

Astrid and Edith: Both "attended private school from six to fourteen" (pp. 261, 264).

"truant together" from school, had virtually identical and unimpressive IQs (of 66 and 67)—for hereditarians presumably attributable more to their bad genes than to their mutual truancies. Jessie and Winifred, who were brought up virtually next door to one another and who "gravitated to one another at school at the age of 5," did not differ at all in their IQs (both had transformed scores of 83). Little difference in IQ and little difference in social environment existed for Molly and Dorothy (whose IQs were 99 and 95), Joan and Dinah (77 and 74), and Joanna and Isobel (123 and 125). Were Joanna and Isobel's IQs attributable more to impressive genes or to the fact that both had strict upbringings (in biologically related families), were required to read and study a lot as children, and attended a prestigious private school together?

Altogether, 14 of Shields's 37 twin pairs revealed such strong similarities in their social environments. Thirteen additional pairs, who did not attend the same school but clearly revealed other similarities in their social environments, were classified as "possible" in similarity of social environment. Thus over two-thirds (27 out of 37) of Shields's twin pairs were clearly raised in similar social environments. These twin pairs differed by an average of only 5 IQ points (see Table 3.4), and their IQ correlation of .89 is only slightly lower than the IQ correlation commonly found in studies of monozygotic twins raised *together.*

The IQ correlation for the remaining 10 pairs who were classified as "minimal" in social environmental similarity, is much lower: .45 for raw scores and .44 for transformed scores. Despite the smallness of the samples, these correlation coefficients differ significantly from the .89 correlation characterizing those with strong or possible environmental similarity ($P < .05$). Again the IQs of the minimally similar pairs differ on the average almost as much as do the IQs of hypothetical pairs of totally unrelated individuals from a general population (14.7 IQ points). Granting the problem of the smallness of the subsample, that correlation of .44 (or .45) is a fairer estimate of the statistic r_{MZA} than is the IQ correlation of .77 commonly cited for all 37 of the Shields twins. And yet, even that correlation of .44 (or .45), based on 10 pairs, is itself inflated: Five of these 10 pairs were reunited for a time during childhood, and 3 of those 5 were raised in related families.

It is instructive to consult the Shields appendix to get some idea of how at least reasonably separated identical twins might look. A good example is the pair of Marjorie and Norah (Shields, 1962, pp. 203–6), tested at age 36, who revealed the greatest IQ discrepancy of the entire Shields sample: 80 (Marjorie) and 50 (Norah) for raw scores, corresponding to 111 and 87 respective transformed points. This is a difference of 24 IQ points, almost two full standard deviations. Marjorie and Norah had *widely* differing social environments: They were separated at 1 year, 10 months; they never even knew they were twins until they were 19 years old; they were never reunited in childhood; and they were raised by unrelated foster families. Marjorie, whose transformed score was 111, clearly had the more advantageous social environment: She was raised by a doctor and attended private schools and one "well-known academy." Norah, whose transformed IQ was 87, was raised by a decidedly unsuccessful unfrocked itinerant evangelical minister.

There are of course other examples from the Shields appendix (10 altogether) of IQ discrepancies that accompany discrepancies in social environment. In each case, the twin with the more advantageous environment had the higher IQ score. For example, one twin, Herta, was raised with little schooling while her twin sister Berta was raised by a doctor and given schooling beyond high school (pp. 221–22). Not surprisingly, Berta's IQ came out to be quite a bit higher (20 transformed points) than Herta's. Similar stories describe Russell and Tristram, who differed by 19 points (pp. 165–68); Jacqueline and Beryl, who differed by 14 points (pp. 217–18); and Olwen and Gladys, who differed by 12 points (pp. 228–29).

In the Newman et al. study, the contrast between the environmentally similar and the relatively more separate pairs is even more striking than in the Shields sample. At least 4 of Newman et al.'s pairs revealed strong similarities in social environment, and 8 more revealed moderate similarities in social environment (Table 3.4). Edwin and Fred, tested by Newman et al. at age 26, had virtually identical Stanford-Binet IQs of 91 and 90 respectively. They are treated in the hereditarian literature as being separately raised, despite the fact that both lived "in the same New England town," and "even went to the same school." They were raised in families "of essentially the same social and economic status." They have both "been electricians for telephone companies," and they "lay stress on the fact that each owns a fox terrier dog named Trixie" (Newman et al., 1937, p. 281). "On the whole, there seem to be no marked differences in their social environments" (Newman et al., 1937, p. 283). One might, however, suspect Newman et al. of naiveté concerning this "separated" pair since on a preceding page, they state that Edwin and Fred "went to the same school for a time but never knew that they were twin brothers" (p. 281). Perhaps the chance for Edwin and Fred to attend the Chicago Fair generated some inventive definitions on their part of being "raised separately"; they appear in Table 3.4 as having "strong" environmental similarity.

So do Kenneth and Jerry, Harold and Holden, and Ada and Ida, all described in Table 3.5. Eight additional pairs in the Newman et al. study showed marked environmental similarities other than having attended the same school and are classified as "possible" in environmental similarity. Three are described in Table 3.5: Maxine and Virginia, Thelma and Zelma, and Edith and Fay. Altogether, then, 12 of Newman et al.'s 19 "separated" twin pairs revealed clear similarities in educational and social environments. These 12 pairs differed in their intrapair IQs by an average of about 4 Stanford-Binet IQ points (Table 3.4), and their IQ correlation is .91, which roughly equals the correlation commonly found in studies of identical twins raised together. (In fact, Newman et al.'s sample of 50 MZ pairs raised together showed an identical IQ correlation of .91.) Again one must ask how separated the "separated" identical twins studied by those who seek to estimate the genetic heritability of human IQ actually were.

The story is quite different for the remaining 7 pairs, classified as "minimal" in social environmental similarity in Table 3.4. They differ by an average of nearly 16 IQ points (15.57), once again revealing the tendency for the more separated twins to show an average intrapair IQ difference that approaches the theoretical value of 17 points for the average difference between pairs of biologically unrelated individuals in a general population. The IQ correlation for these 7 pairs is only .36 and differs significantly, despite the small subsample sizes, from the .91 correlation for the "strongs" and the "possibles" combined ($P < .05$). (For 6 pairs considered by Newman et al. themselves to be "large" in educational differences, pp. 327–28, the IQ correlation can be calculated as .44.)

The estimate of r_{MZA} from the Newman et al. data, with the effects of social environment removed as much as possible, is thus .36. This correlation is only suggestive, of course, given the small number of pairs upon which it is based. Even this correlation, though, low as it is, may itself well be inflated by the sampling bias for Newman et al.'s separated pairs already discussed. Furthermore, of the 7 pairs upon which the .36 correlation is based, 1 was reunited for a time during the teenage years, and 3 others were raised in related families. Clearly, an estimate of h^2 based on the more separated among Newman et al.'s "separated" pairs would be less than .36. That is a long way from Jensen-like estimates of .80.

These 7 pairs with no obvious intrapair social environmental similarities afford one an opportunity to observe contrasting social environments along with contrasting IQs. Newman et al.'s famous pair, Gladys and Helen, differed by 24 Stanford-Binet IQ points (92 and 116 respectively) when tested at the age of 35. This IQ difference is sometimes cited as the largest IQ difference in all four of the separated twin studies (including Burt's), although Marjorie and Norah in the Shields sample also differ by 24 points if transformed scores are used. It comes as no surprise that the separation in social environments of Gladys and Helen, especially their educational environments, was also among the greatest in the twin studies. Helen had the higher IQ and by far the more advantageous social environment: She completed high school and "took a Bachelor's degree from a good Michigan college and soon afterward became a schoolteacher" (Newman et al., 1937, p. 245). In contrast, Gladys had no formal education beyond the third grade. The investigators state quite correctly that their differences in social environment are "too clear to need emphasis" (p. 255).

Other pairs in the Newman et al. sample similarly show not only a fair IQ discrepancy but a fair discrepancy in social environment as well, prior to being tested for IQ. For example, James completed high school and scored 96 on the Stanford-Binet, while his brother Reece "attended a mountain school when he felt so inclined" (p. 307) and scored 19 points less. Mary (IQ of 106) completed high school; her sister Mabel (IQ of 89) completed only the eighth grade. And so on for a total of 7 of Newman et al.'s pairs.

The effect of similarity in social environment is also evident in the Juel-Nielsen sample, although the differences are not statistically significant presumably because of the small sample size (Table 3.4). It does appear, from the extremely detailed Juel-Nielsen appendix, that the degree of separation for all 12 twin pairs was somewhat more complete than was the case for the other two studies. Pairs having strong or possible similarity in social environment reveal somewhat less IQ difference (6.71 for time₁ scores, and 5.14 for combined scores) than do pairs classified here as relatively less in environmental similarity (8.00 and 8.20). The IQ correlations are higher for the 7 pairs who are relatively more similar in social environment. (The correlation is inflated slightly if time₁ and time₂ test scores are combined, as done by Jensen, 1970b.) Examples of environmentally similar pairs from the Juel-Nielsen study are given in Table 3.5. Environmentally more different pairs (5 in total) are exemplified by Viggo and Oluf, who differed by 12 IQ points at the age of 77 (the largest difference of the twelve). The somewhat brighter Oluf (IQ of 112) attended college and became a research assistant in botanical genetics. His more modestly endowed brother attended school "infrequently" as a child, had no college education, and became a farmer.

If all three studies are combined (the last column in Table 3.4), it becomes apparent that 46 of the MZ twin pairs were raised in educational, socioeconomic, and interpersonal environments that were strongly or moderately similar, which is over two-thirds of the 68 pairs. Yet every one of these 46 pairs is considered separately raised in uncorrelated and different environments by the bulk of the literature on IQ heritability. The averaged IQ correlation for these 46 pairs is .87, close to the commonly observed estimates of the IQ correlation for

MZ twins raised together. One blatantly obvious explanation for this is that the social environments of these 46 "separated" MZ pairs were in reality not much different (only somewhat less similar, at best) from those of the identical twin pairs appearing in the professional literature as raised *together* (MZTs).

The overall picture across the three studies is quite different for the remaining 22 pairs who are characterized by less similarity in social environment. Their averaged correlation is .43, which is significantly different from the averaged correlation of .87 ($P < .001$). This correlation coefficient of .43 might conceivably be used as a direct estimate of h^2 if it were not for the fact that 11 of these 22 pairs were reunited for a time in childhood, raised in related families, or both (see Chapter 3, note 12), and if it were not for other potential sources of correlation inflation such as sample bias, unknown and unmeasured environmental variables, and other factors (covariance; the age-IQ confound; and the fact that the same investigators tested both members of a twin pair) considered in the next section.

On the basis of these analyses one might profitably ask what the IQ correlation is for the 11 MZ pairs who had no similarities in social environment, who were not reunited in childhood, and who were not raised in related families. Three of these come from the Newman et al. study, and 3 from the Juel-Nielsen study (too few to calculate an IQ correlation). That leaves 5 MZ pairs from the Shields study who can be thought of as being truly separately raised, and the IQ correlation for these 5 pairs is only .24 (for both raw and transformed scores).

This correlation of .24 is as conceptually important as it is empirically meager, coming as it does from only 5 pairs. But even with such limited data, the greater the extent to which one removes environmental similarity as a source of the IQ correlation, the lower, by a large margin, is that correlation. The resulting correlation is clearly a long way from Jensen's correlation of .82, based on the separated twins studies, said by him to be an "upper bound" estimate of "the heritability of IQ in the English, Danish, and North American caucasian populations sampled" (1970b, p. 133). On the basis of this reanalysis of the same samples, the "upper bound" of IQ heritability is considerably below where Jensen thought it to be.

Controlling for childhood reunion alone reduced the h^2 estimate (.85 versus .55 averaged). Controlling for only relatedness of adoptive families also reduced it (.77 versus .61). Controlling for similarity in social environment alone reduced it markedly (.87 versus .43). Fur-

thermore, unmeasured environmental similarities may also characterize these twin pairs. Combining all this information (including the .24
correlation based on the 5 completely separated Shields pairs), one is
tempted to make the tentative extrapolation that with larger samples
and a few more environmental variables to control, the IQ correlation
for truly separated identical twin pairs would be extraordinarily low.
Based on the reanalysis of the data from three of the four known studies of separated identical twins, it seems reasonable to suggest that
the IQ correlation characterizing pairs of individuals with absolutely
identical genes and absolutely uncorrelated environments would be
extremely low. If this is in fact true, then the identical genes of monozygotic twins would have little to do with their comparative IQs.

A Note on Measurement Error

In addition to the four variables that contribute to the inflation of the
IQ correlation of separated identical twins already discussed, there
exist two sources of artificial inflation, uncovered by Kamin (1974),
which are classifiable as measurement errors: (1) use of the same examiner to test both members of a twin pair; and (2) faulty age and sex
standardization of the IQ tests used. If the same examiner tests two
biologically related individuals on any attitudinal or behavioral measure, it is always possible that the examiner's biases or expectations,
whether conscious or not, can operate to increase the similarity between the pair members. This can happen even with paper-and-pencil
tests, since the examiner's mannerisms, verbal instructions, and nonverbal cues can potentially affect the examinee's responses.

    In the Newman et al. study, it is not clear whether the Stanford-
Binet test was administered to twin pairs by different examiners or the
same examiner. For the Juel-Nielsen study, however, it is clear that 11
of the 12 pairs were indeed given the Wechsler IQ test by the same
examiner (Juel-Nielsen, 1965, p. 54). Consequently one cannot compare the IQ correlations for pairs tested by the same examiner to the
IQ correlation for pairs tested by different examiners in either study.

    For the Shields sample, however, such a comparison is possible if
one includes in the analysis three pairs that Shields himself did not
include, though the results are only suggestive. Kamin (1974, p. 49)
notes that 5 pairs who were tested by different examiners had an IQ
correlation of only .11, as compared with .84 for those tested by
Shields himself (35 pairs). The subsample is small (5 pairs), and for 3 of
these 5 pairs, Shields reported evidence of brain damage of one or

both members of the pair. (This would contribute to lowering their IQ similarity.) Still, it seems clear that at least some same-examiner effect was present in the Shields data, and thus could well have been one source of the twins' IQ similarity. Same-examiner effects might well have been present in the Newman et al. and Juel-Nielsen studies, also, though same- versus different-examiner effects cannot be compared.

The second issue Kamin raises has highly important implications for IQ testing generally. He found that the IQ scores for Newman et al. (who used the 1916 version of the Stanford-Binet) and Juel-Nielsen (who used a Danish translation of the Wechsler) correlated with the twins' ages. IQ as originally conceptualized is the ratio of "mental age" to chronological age (hence "intelligence quotient"), where mental age is defined by one's test performance relative to the average test performance on the same test items for others of differing ages. A correctly standardized test should therefore show the "average" person having a mental age equivalent to his or her chronological age, that is, an IQ of 100—regardless of age. Hence, in a sample IQ score should not correlate with age.

Newman et al., however, themselves found an age-IQ correlation of −.49 for their 50 nonseparated MZ pairs. They did not report any age-IQ correlation for their separated pairs, but it turns out to be −.22 (Kamin, 1974, p. 57). A negative correlation would mean that the older one is, the stupider one becomes: Hence, the age standardization for both the separated and the nonseparated samples was faulty. As further evidence of faulty standardization of the Stanford-Binet, Kamin notes (1974, p. 63) that a study of nonseparated MZ twins by Babson et al. (1964) showed a positive age-IQ correlation of .59. Furthermore, a study of nonseparated MZs by Willerman and Churchill (1967) reveals an identical age-IQ correlation of .59. Kamin also notes that pairing the Newman et al. separated twins by adjacent ages rather than by twinship results in a correlation of .67 for 7 male pairs (1974, p. 60). (Loehlin et al., 1975, p. 295, point out that the 12 female Newman et al. pairs correlate only −.02 when paired by age, which suggests some confounding of sex as well as age.) The 9 female pairs in the Juel-Nielsen sample correlated .59 in their IQs when paired by age. Both bits of evidence taken together (linear age-IQ correlation and IQ correlation when paired by age) suggest that for twins, Stanford-Binet and Wechsler IQ may be predictable by age as well as by twinship. Because MZ twins are of identical age, a potential source of IQ similarity, in addition to genetic identity, is introduced.

JENCKS ET AL.'S PATH ESTIMATES USING
SEPARATED IDENTICAL TWINS

Path analysis models of IQ heritability are introduced in Chapter 4, but it is worthwhile at this juncture to examine closely the path model and analysis used by Jencks et al. (1972, pp. 309–15) in their study of separated identical twins. If, as suggested in the foregoing analysis, a true observed IQ correlation for separated identical twins is somewhere in the neighborhood of .20 to .40, then extremely low values for h^2 are implied. The IQ correlation for separated twins cannot itself be used as a direct estimate of h^2, since if there is any remaining environmental similarity after separation (that is, if the correlation r_{E1E2} is greater than zero) or if there exists any remaining dependence of environmental similarity upon the twins' genetic identity (that is, if there is any nonzero and positive gene-environment correlation or covariance), then the twins' IQ correlation (r_{MZA}) will necessarily be greater than h^2. This would mean of course that the IQ correlation by itself overestimates h^2, yet Jensen, Herrnstein, and Eysenck have advocated using the separated twin correlation itself as a direct estimate of h^2.

To investigate the h^2 values one might obtain from different combinations of hypothetical values for environmental correlation and gene-environment covariance, I use Jencks et al.'s path model and parameter values. The Jencks et al. model appears here as Figure 3.1.

Jencks et al. examined the data from all four of the separated twins studies, arriving at an averaged IQ correlation of .81. For the path analysis, however, they used an estimate of .75, based on only the Newman et al. study, correcting the original Newman et al. correlation (.67) for attenuation. Such values considerably overestimate the observed correlation for separated MZ twins, and I therefore use smaller values. Many investigators besides Jencks et al. persist in using grossly inflated values for this observed IQ correlation. Fulker, for example, uses the absurdly high .95 (1975, p. 519),[14] Rao et al. use .69 (1976, and from Rao et al., 1974, p. 353), and Rao and Morton use .679 (1978, p. 179).

By using estimates between .20 and .40 for the observed IQ correlation (abbreviated as $r_{Y1Y2(MZA)}$), it is being argued that IQ correlations for MZ twins raised in reasonably separate educational, socioeconomic, and interpersonal environments are roughly not more than .40 nor less than .20. It is to be stressed that these estimates, which are based on the foregoing analyses, particularly the analysis of similarity in social environment, are rough guesses at best.

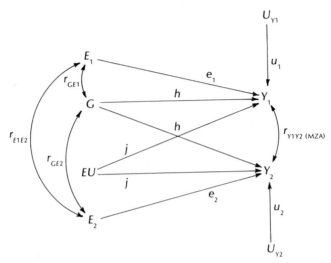

Figure 3.1. The Jencks et al. path model for separated identical twins. *Source:* Jencks et al., 1972, p. 312. *Note:* E_1 and E_2 are Jencks et al.'s EF_1 and EF_2 (for family environments); U_{Y1} and U_{Y2} are Jencks et al.'s ER_1 and ER_2 (for nonfamily or residual environments); the correlation coefficients r_{E1E2}, r_{GE1}, and r_{GE2} are Jencks et al.'s k, a_1, and a_2 respectively. The construct EU represents intrauterine environment. Jencks et al. assume that $e_1 = e_2 = e$; and that $u_1 = u_2 = u$.

Variables (constructs) E_1 and E_2 represent the family environments of any given separated twin pair. Thus r_{E1E2} is the degree of environmental similarity remaining after their separation. Correlations r_{GE1} and r_{GE2} are the gene-environment correlations for each pair member; Jencks et al. do not assume these are necessarily equal. G is represented as one variable, since for MZ twins, $r_{G1G2} = 1$. The correlation $r_{Y1Y2(MZA)}$ is the actual observed pair-correlation of the IQs of the twins. Four path coefficients (standardized regression coefficients) appear in the model: h (h^2 is heritability); e (e^2 is environmentability; Jencks et al. arbitrarily assume that $e_1 = e_2$); u (the effect of residual or nonfamily environment; it is arbitrarily assumed that $u_1 = u_2$); and j (the effect of intrauterine environment, estimated by Jencks et al. on the basis of the observed relationship between birth-weight differences in MZ twins and their IQ differences).

The model in Figure 3.1 yields the following path estimation equation:[15]

$$r_{Y1Y2(MZA)} = e^2 r_{E1E2} + h^2 + j^2 + her_{GE1} + her_{GE2},\qquad(3.2)$$

which is Jencks et al.'s equation *(42)* (1972, p. 312). The IQ correlation $r_{Y1Y2(MZA)}$ is thus seen as arising from four sources: remaining environmental correlation after separation ($e^2 r_{E1E2}$); heritability (h^2); intrauterine effects (j^2); and gene-environment correlation ($her_{GE1} + her_{GE2}$). Since Jencks et al. explicitly partition the effects of social environment from those of the intrauterine environment, the overall effect "of environment" is not e^2 but ($e^2 + j^2$).

Before proceeding further, it will be instructive to point out some of the assumptions (largely arbitrary) in the Jencks et al. model. Some are more explicit than others, and several are highly implausible.

1. It is assumed that residual variables U_{Y1} and U_{Y2} (representing nonfamily environments) are uncorrelated with each other and that each is also uncorrelated with E_1, E_2, G, and EU. This amounts to arbitrarily and implicitly setting at zero 9 correlations: (the 4 between U_{Y1} and each of the exogenous variables E_1, E_2, G, and EU; the 4 between U_{Y2} and the exogenous variables; and the 1 between U_{Y1} and U_{Y2}). Yet surely at least some of these correlations are in reality nonzero. For example, one might justifiably suspect at least some nonzero correlation between family (E) and nonfamily (U_Y) environments. Or, to take another example, it seems absurd to assume that the twins' nonfamily environments are completely uncorrelated after their separation. (To define as do Jencks et al. the U_Y as environments "specific" to each single pair member, thus by definition uncorrelated with each other, is an obfuscation. This noncorrelation by definition of nonfamily environments, that is, residual variables, is an approach used in other path analyses as well, notably those of Rao et al., 1976; Rao and Morton, 1978; and Loehlin, 1978, considered in Chapter 5.) Yet another example: Is it really plausible to assume, as Jencks et al. do, that G and the residual environments U_{Y1} and U_{Y2} are independent, yet proceed, as Jencks et al. do, with the argument that G and the family (common) environments (E_1, E_2) are correlated? Surely people who look exactly alike tend to get treated alike in their nonfamily environments even if they are reasonably well, though not perfectly, separated. If any 1 of these 9 correlations is indeed nonzero, then it would have to enter equation *(3.2)*, which would mean that the equation would be misspecified as stated.

2. It is assumed that EU is uncorrelated with E_1, E_2, and G, and these correlations are set at zero. But: is it really plausible to assume, as do Jencks et al., that the socioeconomic status of the mother (subsumed under E) has no nutritional or chemical effects on the intrauterine environment? Why persist with the far-fetched argument that intrauterine environment affects IQ and yet assume that the social environment has no effect on the intrauterine environment?

3. The model, and thus equation (3.2), assumes recursiveness, or one-way causation. This means that no allowance is made, for example, for the effect of a twin's IQ score upon how he is treated in his own environment—such as in school! This assumption is of course grossly implausible.

4. The model does not allow for the possibility of sample bias, especially biases of the sort that might result in the underrepresentation of environmentally separated pairs. This kind of bias could well have been particularly severe with the Newman et al. study, which provides the basis of Jencks et al.'s IQ correlation.

5. The assumption is made that there is no direct causation (no single-headed arrow) from Y_1 to Y_2, nor from Y_2 to Y_1, an assumption that comes up again in other studies considered later. Thus Jencks et al. are in effect assuming that

$$\beta_{Y1Y2} = \beta_{Y2Y1} = 0,$$

in other words, that the measured IQ of one twin has no direct causal effect on the measured IQ of the other twin, a clearly implausible assumption. Although Jencks et al. note this possible effect for raised-together kinships, they do not consider its implications for raised-apart MZ twins (1972, p. 296). Since many of these twins (over half of Newman et al.'s) probably had a fair amount of personal contact with one another, any one pair member's IQ could have directly influenced that of the other through their interaction. If that has been the case, then quantities for either β_{Y1Y2} or β_{Y2Y1} (or both if nonrecursiveness, or two-way causation, is suspected) would have to enter equation (3.2). The equation would thus be misspecified as stated.

If for the sake of argument, one takes these assumptions as given (including the assumptions of additivity, linearity, and no nonrandom measurement error), it is possible to calculate illustrative values of h^2 from equation (3.2) for alternative values of e^2, r_{E1E2}, j^2, and $her_{GE1} + her_{GE2}$. Table 3.6 shows the results for three values of $r_{Y1Y2(MZA)}$, namely .40, .30, and .20.

These estimates of the IQ correlation ranging from .20 to .40 are

Table 3.6. **Heritability Estimates Using Three Alternative Values for the Observed IQ Correlation for MZAs**

$(her_{GE1} + her_{GE2})$	e^2	r_{E1E2}	h^2 if observed IQ correlation is		
			.40	.30	.20
.00	.00	.04	.34	.24	.14
		.20	.34	.24	.14
	.35	.04	.33	.23	.13
		.20	.27	.17	.07
.12	.00	.04	.22	.12	.02
		.20	.22	.12	.02
	.35	.04	.21	.11	.01
		.20	.15	.05	<0
.20	.00	.04	.14	.04	<0
		.20	.14	.04	<0
	.35	.04	.13	.03	<0
		.20	.07	<0	<0

Note: Using equation (3.2), where $j^2 = .06$.

based on MZ twins who were at least reasonably well separated in their social environments, yet certain environmental similarities remained. For example, of the 22 pairs categorized as "minimal" in similarity of social environment (whose averaged IQ correlation = .43), half were reunited in childhood, raised in related families, or both (Table 3.4 and note 12). In terms of Jencks et al.'s model, this means that even after removal of a certain amount of environmental similarity, there nevertheless clearly remains some degree of correlation of environments. Thus, it is safe to assume as do Jencks et al. for separated MZ twins that $r_{E1E2} > 0$, even for reasonably separated pairs.

I employ two values for r_{E1E2}, the lowest and the highest permitted by Jencks et al. On the basis of studies of adopted children (Burks, 1928; Leahy, 1932, 1935; Skodak and Skeels, 1949), Jencks et al. (1972, pp. 278, 313) find the averaged correlation between the educations of natural and adoptive parents to be about .20. (In a more recent study, Scarr and Weinberg, 1976, p. 734, similarly found that the education correlation of natural and adopting mothers was .22.) This is in effect an index of selective placement of the offspring with respect to education. Assuming that parental education is a reasonable index of certain aspects of environment, Jencks et al. use this .20 correlation as a maximal value for separated twins, based on the form of separation in which one twin is raised by the natural parents and the other is adopted. In cases in which both members of the pair are adopted, the averaged correlation was estimated by Jencks et al. to be about .04. Accordingly I also assume here that r_{E1E2} is between .04 and .20. This is conservative, as .04 is exceptionally low, even for reasonably separated twins (noted by Jencks et al., 1972, p. 312).

For the gene-environment covariance ($her_{GE1} + her_{GE2}$), I employ three values. For a maximal value, I use the Jencks et al. overall estimate of $2her_{GE}$ in a general population of individuals, where individuals are raised by their natural parents. If MZ twins who have been reasonably separated in fact reflect no more and no less gene-environment correlation than such a general population, then according to Jencks et al.,

$$her_{GE1} + her_{GE2} = 2her_{GE} = .20.$$

For a minimal value, I estimate her_{GE1} and her_{GE2} separately. Let the subscript (1) represent a twin who remains with the natural parent, the subscript (2) a twin who is adopted. Following Jencks et al., her_{GE2} is assumed to be less than her_{GE1}. Thus $her_{GE1} = her_{GE} = .10$, and $her_{GE1} > her_{GE2}$.

For her_{GE2}, Jencks et al. (1972, p. 313) employ the following estimate, based on an analysis of parent-offspring correlations:

$$her_{GE2} = .16her_{GE} \sqrt{1 + p}, \tag{3.3}$$

where p represents Jencks et al.'s estimated correlation between the genotypes (Gs) of both parents of an individual (and is thus a correlation for assortative mating) and which based on Jencks et al.'s Table A-5 (p. 281), can be calculated to be between .51 and .23. From equation (3.3), her_{GE2} is between .019 and .017; or roughly .02. The overall minimal estimate of ($her_{GE1} + her_{GE2}$) is thus .10 + .02 or .12. The gene-environment covariance quantity is thus between .12 and .20 (Table 3.6).

On the argument that by using only the reasonably well separated MZ pairs any dependence of environmental similarity upon genetic similarity has been removed, I assume a third value, a value of zero, for $her_{GE1} + her_{GE2}$ in Table 3.6, being as conservative as possible with respect to the allowable amount of gene-environment covariance.

I need now estimate only e^2 and j^2 in equation (3.2) to derive alternative values for h^2. Jencks et al. (1972, p. 313), arguing that a twin's birth weight affects his or her IQ and that intrauterine environment affects birth weight (such that the heavier of two MZ twins has the higher IQ), estimate that $j^2 = .06$ (thus $j = .25$), the value I employ. Some have argued convincingly that birth weight has virtually no effect on IQ (e.g. Kamin, 1974, pp. 161–73; Kamin, 1978), but for the sake of argument I stick with the Jencks et al. estimate. Finally, I use a maximal value of .35 for e^2, based on Jencks et al.'s overall estimate of e^2. Again being as conservative as possible, I use zero as the minimal value of e^2.

The h^2 values resulting from equation *(3.2)* can now be seen. Comparing rows in Table 3.6, if one assumes that the IQ correlation of reasonably separated twins ($r_{Y1Y2(MZA)}$) is as high as .40, then under the assumptions and estimates used, h^2 can be no greater than .34 nor less than .07. If $r_{Y1Y2(MZA)}$ is in the neighborhood of .30, then h^2 falls anywhere from .24 to zero. If $r_{Y1Y2(MZA)}$ is roughly .20, then the maximal value of h^2 is only .14, and the minimal values are less than zero. Comparing columns, for the minimal estimates of ($her_{GE1} + her_{GE2}$), e^2, and r_{E1E2}, h^2 ranges from .34 to .14. For maximal estimates, h^2 ranges from a maximum of .07 to less than zero.

The most reasonable and general interpretation for these results is that if the correlation between the IQs of monozygotic twins who are separately raised is anywhere from about .40 to about .20, then the percentage of the total variance in the measured IQ of either twin that is explained by their intelligence genotype is somewhere between 34 and zero. This is a long way from Jensen-like estimates of 80 percent, and from other common estimates of about 60 to 70 percent. This conclusion also differs considerably from that of Jencks et al.; they estimated (on the basis of separated twins) that h^2 was between .47 and .55 (1972, p. 314). If more plausible values than Jencks et al.'s .75 are used for the observed IQ correlation, the model, even given its questionable assumptions, is compatible with the hypothesis that the heritability of human IQ is zero or nearly zero, which does not of course mean that it is zero, or even likely to be. On the basis of a path analysis of separated monozygotic twins, no compelling reason appears to postulate the existence of any genes "for" intelligence.

SUMMARY

Identical twins regarded in the professional literature as being separated early in life and raised apart were not actually so raised. The IQ correlation of separated identical twins, often presumed by Jensen and others to be a direct estimate of IQ heritability, can be artificially inflated by: (1) sample bias in the original studies; (2) reunion of the twins after their initial separation but prior to being tested for IQ; (3) upbringings in families that were related through friendship or kinship; and (4) upbringings in highly similar educational and social environments. Other sources of artificial inflation of this correlation include the use of (5) transformed IQ scores (as Jensen did with Shields's scores); (6) combined time$_1$ and time$_2$ scores (as Jensen did with Juel-Nielsen's scores); and (7) Jensen's ill-conceived "difference correlation." Furthermore artificial inflation of the IQ correlation can result from (8) the use of the same examiner to test both members of a twin pair, and (9) faulty standardization of IQ tests.

About two-thirds of the 68 identical twin pairs originally studied by Shields, Newman et al., and Juel-Nielsen do not fit any reasonable definition of being raised separately in uncorrelated environments. They were reunited for a time in their youth prior to testing, or raised in clearly related families, or raised in common social and educational environments, or some combination of these. Jensen, Herrnstein, Eysenck, and most of the professional literature on IQ heritability, have consistently treated twins as though they were raised in different environments. They were not.

Introductory texts in the otherwise environmentalist field of sociology continue to reproduce the supposed finding that separated monozygotic twins correlate highly in their IQs (see for example CRM's *Society Today*, 1973, pp. 86–87; Fernandez, 1977, p. 31; and Cole, 1979, p. 93). An even greater number of introductory texts in the fields of psychology and education perpetuate this false belief. Even texts in genetics do so (for example, McClearn and DeFries, 1973). Perhaps among the most eminent of geneticists is Theodosius Dobzhansky, a world-renowned Russian-born geneticist who came to the United States as a young man to escape the environmentalism of Stalin's favorite geneticist, T. D. Lysenko. In a book published shortly before his death in 1975, Dobzhansky (1973), citing all four of the separated twin studies, appeared unaware of the possibilities of the pronounced environmental similarities among separated twins, arguing quite flatly that their IQ correlation "is a matter of genetics" (p. 13).

The IQ correlation for the not so separated among the "separated" MZ twins is indeed quite high. In contrast, the IQ correlation is quite low for subsamples of these twins who were in fact reasonably separated according to the criteria of no reunions, only moderate family relatedness, and minimal similarity in social environment. Borrowing a path model and parameter values from Jencks et al. (1972), using revised values for the observed IQ correlation for separated twins, and allowing for remaining environmental correlation after separation and remaining gene-environment covariance, *low* heritability values (down to zero) are obtained.

Yet Jensen continues to insist that "the fact that MZ twins reared apart have much more similar [more highly correlated] IQs than DZ twins raised together *leaves really no doubt* of the heritability of IQ" (1975, p. 177, italics added; cf. Jensen, 1978b). But the IQ correlation for environmentally separated MZ twins is the *same or less* than the oft-cited correlation of .50 to .60 that characterizes fraternal twins raised together. Even some researchers of an environmentalist persuasion (such as Bronfenbrenner, 1975, p. 121) have erroneously argued that MZs raised apart correlate more highly than DZs raised together. In fact, the mean intrapair IQ difference for reasonably separated MZ twins tends to approach that of a general population of biologically unrelated individuals reared apart.

All this is powerful evidence for the extensive operation of the social environment upon the IQ similarity of "separated" monozygotic twins and weak evidence for the operation of genes. It points to the conclusion that the heritability of IQ is not an estimable quantity. But if one nonetheless attempts its estimation despite the implausible assumptions necessary to do so, then concluding in favor of low heritability values is plausible. In sum, given the available methods and data, there once again appear to be no compelling reasons to postulate the existence of any genes "for" intelligence.

Chapter 4

Studying Kinships
Two at a Time

In any scientific inquiry, one proceeds as if certain factors were given, making assumptions about unknown factors. To arrive at a numerical estimate for h^2 for human IQ for a given population, quite a few factors must be taken as given. If these givens cannot be demonstrated empirically, they must be assumed a priori, and the plausibility of these assumptions must be considered. For an analysis of IQ heritability consideration of the assumptions is not merely an academic exercise, given the policy implications of such research. This chapter is devoted to the assumptions of heritability analysis, discussing the arbitrariness and implausibility of many of them.

THE REQUIREMENTS OF HERITABILITY
ESTIMATION

Estimation of h^2 for IQ even for a single population or group involves many assumptions. Some of these, for example, the requirement of equal intrapair environmental similarity when comparing kinships, the requirement of uncorrelated environments when studying separated MZ twins, have already been touched on, but I examine them in more detail in this chapter. I begin with what might be called the standard requirements of multivariate analysis. Later I undertake a more technical examination of requirements specific to the models used in the literature on IQ heritability, particularly the (often only implicit) requirements for the models of Jensen, Herrnstein, and Eysenck.

To show that measured IQ for a given, specified population has a nonzero and quantitatively estimable genetic heritability, one must necessarily demonstrate, or failing that, assume, the following:[1]

1. *That there is low (preferably zero) measurement error.* Intelligence is an abstract, hypothetical concept. The degree to which it is actually measured by standard IQ tests is the degree of measurement error of the IQ test in question. Measurement error subsumes both validity (does the test measure what it is supposed to measure?) and reliability (the stability of an individual's scores over time and with different researchers or different forms of the same test). According to modern measurement theory (e.g., Kerlinger, 1973), reliability subsumes validity so that a measure can be both valid and reliable, or reliable and invalid, but if it is unreliable, it is necessarily invalid. Standard IQ tests such as the Stanford-Binet or the Wechsler have indeed been shown to be reasonably reliable, but whether or not they constitute a valid measure of intelligence is a long-standing and unresolved debate. Recent evidence exists that IQ is reliable only for short time spans and can be markedly changed through special education programs (McCall et al., 1973).

Without entering in detail into all aspects of this debate, I will later note four clear flaws in the hereditarian argument that fall under the heading of measurement error: (a) The question of whether or not IQ tests constitute as valid a measure of intelligence for minorities, particularly black persons, as they do for the white, middle-class populations for which they were designed; (b) the question of whether or not intelligence is unidimensional, that is, having only one principal component or factor, g (for "general intelligence") as opposed to a number of dimensions or factors; (c) the adequacy of using the popular correction for attenuation procedure on IQ scores; and (d) inadequate age and sex standardization of test scores.

2. *That there is low (preferably zero) sample bias.* In order to infer that any h^2 calculated on a given sample of persons approximates h^2 in a population from which the sample was drawn, the researcher must show that the sample in question is indeed a reasonable representation of that population. If in fact a given sample differs from its hypothetical population with regard to the variables under study, then some degree of sample bias exists. Sample sizes in the majority of kinship studies through about 1975 tended to be quite small (averaging around 35 pairs, according to Lewontin, 1975), thus causing one to question their representativeness with respect to their populations. I have already reviewed evidence that the "separated" identical twins are probably not accurate approximations of their populations.

3. *Ceteris paribus for the environment.* The principle of "all else constant" is the backbone of the scientific method. To show that G

affects IQ, one must show that people who differ in their hypothetical amounts of G also differ in measured IQ even if they are identical on E. This is what is meant by "holding environment constant." To show that individuals who are the same on G (such as MZ twins) are also the same in their IQs, the environments of the individuals in question must be completely random. (This is essentially the strategy involved in attempting to estimate h^2 from r_{MZA}.) If people could be treated like fruit flies or rats, it would be possible, via experimental designs, to control the environment and at the same time systematically manipulate genotypes over several generations by selective breeding. But in the analysis of human IQ scores, one must approximate ceteris paribus through nonexperimental procedures. The validity of the inference that differences in some hypothetical G cause differences in IQ rests upon how closely this condition has been met. That translates into assuming equal environmental similarity across kinships and assuming zero environmental similarity for separated identical twins. (Alternatively, one must actually assign some empirical values to the environmental correlations, or failing that, assume certain inequalities between certain kinships with respect to residual environment.)

4. *That gene-environment interaction is zero (or nonzero but somehow estimable)*. If the effect of G upon IQ is different in magnitude or form for different types of environments (different "values" of E), then gene-environment interaction is said to exist. If not, then additivity, or zero interaction, exists. Virtually all treatments of IQ heritability, most particularly those of Jensen, Herrnstein, and Eysenck, and including Jencks et al. assume additivity.[2] If any gene-environment interaction upon IQ exists, then additional unknowns (representing the IQ differences explained by interaction) are introduced into the system for estimating h^2. (I consider the hypothesis of interaction later.) It bears emphasizing that as the gap between the number of unknowns and the data necessary to solve for these unknowns widens, crucial unknowns like h^2 become less estimable.

5. *That gene-environment correlation is zero (or nonzero but somehow estimable)*. Under the assumption that some G "for" intelligence indeed exists, this G may affect some part of E (social environment) which in turn can affect IQ. There thus may be a nonzero correlation of E and G. Recall that the argument supporting gene-environment correlation runs something like this: (a) People with "intelligent" genes may select advantageous aspects of their environments, and as a result of these environmental advantages, score well on standard IQ tests. (b) People with "intelligent" genes might be given more en-

vironmental advantages by their parents, teachers, or employers, and as a consequence, do well on IQ tests. (c) People similar on G may be treated alike in society, causing them to be more similar on E and thus in IQ. People who are more alike genetically because of genes not "for" intelligence (as, for example, with respect to physical appearance) are also likely to get treated similarly in their environments—and thus perform similarly on IQ tests (see Chapter 2). This last source can show up in heritability analysis as nonzero gene-environment correlation (and as already noted, gene-environment interaction may also have this dual character). This last source is not empirically distinguishable from the previous kind, even assuming that genes "for" intelligence exist, and it can produce artificial inflation of the h^2 estimate. Jensen, Herrnstein, and Eysenck often implicitly assume that all forms of gene-environment correlation are zero. Or, they (especially Jensen) assume that it is nonzero but equal across different kinds of kinships. If any form of gene-environment correlation is indeed nonzero, then yet additional unknowns will enter into the system for estimating h^2 for IQ.

6. *That environmental variance and residual variance are treated separately.* An unfortunate practice in heritability analysis is to assign any variance in IQ that is not attributable to the presumed G to $(1 - h^2)$, that is, to everything else, with "the environment" often used synonymously with "everything else." (Such a procedure leads to overestimation, not underestimation, of environmental effects.) This is a questionable—at best, sloppy—procedure. Following the usual convention, even if E is intended to subsume all conceivable environmental effects (effects of prenatal environment, schooling, peer-group influences, nutrition, and so on) certain effects are by definition not subsumable under the construct E. Such effects would then fall under residual (error) variance, but only if the model explicitly includes a quantity for this residual.

Failure to include a residual term has evidently been a relatively long-standing problem in heritability analysis of IQ scores (as noted by Hogarth, 1974, p. 4). In some of his work, Jensen clearly fails to include such a residual in his models for MZ twins, DZ twins, and unpaired individuals (e.g., 1967). Such residual effects would include variance in IQ resulting from nonlinearity, sampling bias (if the total population variance is being estimated), measurement error not captured by the attenuation correction procedure (such as nonrandom measurement bias, which can arise when IQ tests are given to racial and ethnic minorities), gene-environment interaction (if additivity is assumed), gene-environment covariance (if zero gene-environment correlation is assumed), and other kinds of covariances as well (covariance between environment and interaction, or between genotype and interaction). (Some models, such as those of Jencks et al. and the Honolulu research to be reviewed in Chapter 5, also include specific or nonfamily environment in the residual term.) Unless the distinction between environmental and residual effects is made, the relative explanatory power of genes (h^2) and environment (e^2) cannot of course be legitimately contrasted.

7. *That other sources of IQ variance are separately partitioned.* If one wishes to contrast the various potential sources of IQ variance with each other, then it is necessary to partition these sources. If the only sources of IQ variance in one's model are the constructs G and E, then it is advisable to include a residual source. If for example one explicitly includes a separate term (or terms) for gene-environment interaction, this source of IQ variance is removed from the residual and becomes a separate specified source of IQ variance.

One often finds in the heritability literature that sources of IQ variance are not even conceptually partitioned. A good example is the flagrant error of considering gene-environment interaction as part of "genetic" variance (Herrnstein, 1973, p. 177). Another example is the classification of gene-environment covariance under genetic variance (for example, Jensen, 1973, p. 368). In this case, h^2 would have to be defined as including IQ variance attributable to both G and any G, E correlation. The logic of this seems to be that if in fact variation in G causes variation in part of E which in turn causes variation in IQ, then this effect should be attributable to genetic variance, and thus to heritability. The inadvisability of this procedure has already been pointed out. Sewell Wright (1931) was the first to attempt to partition separately gene-environment covariance in the study of IQ (as pointed out by Morton, 1972), and Jencks et al.'s attempt to do this has already been noted. Other such analyses are considered later.

8. *That all relationships between variables are linear.* Heritability estimation requires that all relationships be linear, since coefficients such as h^2 and e^2 are by definition linear coefficients. But it is impossible to demonstrate the existence (or nonexistence) of linearity, since both G and E, being constructs, are not measured. If any conceivable relationship between G and IQ is appreciably nonlinear, then virtually all procedures used to date in heritability estimation are inapplicable.

9. *That any effects of genotype or environment on IQ are one way (recursive).* A set of variables are *recursively* related when, for all pairs

of variables, if any X_i is presumed to cause X_j, then X_j cannot cause X_i, either directly or indirectly (that is, through other intervening variables). In all analyses of IQ heritability considered in this book, it is assumed a priori that while a person's genotype and environment can affect his IQ, his IQ score cannot affect his environment. (That it cannot affect his genotype is of course taken as axiomatic.) But this assumption is implausible on its face. Clearly, one's IQ score affects how one is treated in one's family, in school, on the job, and so on—that is, one's IQ score clearly affects one's environment. But the IQ heritability researchers must necessarily assume that any causation from genotype to IQ and from environment to IQ is one way, or there will be too many unknowns in the overall estimation system even to consider an empirical solution.[3]

10. *That residual variables are uncorrelated.* Yet another requirement of heritability analysis concerns the "unexplained variance" in IQ that remains after both G and E have explained what they can of the total variance in IQ. Such unexplained variance is subsumable under a residual variable, which must be assumed to be uncorrelated with both G and E; otherwise, two additional unknown quantities (the relationship of G and E to the residual) are inadvertently introduced into the system.

In the analysis of kinship pairs, the IQ scores of both members of a pair are treated as separate variables. In this case, each IQ score will have its own residual variable. If these residuals are correlated with other variables in the model (that is, with G and E), or if they are correlated with one another, additional correlations, and therefore additional unknowns, are thereby introduced. Should such correlations be nonzero, this would greatly increase the number of unknown quantities in the entire estimation procedure, thus rendering quantities such as h^2 and e^2 even less estimable.

11. *That no specification errors appear in the equations.* One kind of specification error exists if the wrong equation or wrong model, that is, an equation or model other than what is clearly and explicitly intended, is used. Another kind occurs when a term or quantity that via conceptual argument should be in an equation or set of equations is not. Loehlin et al. (1975, p. 301) discovered an error of the first kind in one of the Jencks et al. (1972) equations for the correlation between offspring's IQ and the IQ of the adoptive parent (Jencks et al. equation [12], p. 280), which, when corrected according to Loehlin et al. resulted in an h^2 of .60 (rather than Jencks et al.'s overall estimate of .45). In a later article, Loehlin (1978) located another specification er-

ror in Jencks et al. that, he argued, roughly offset the first. Goldberger (1978b, 1978c, and 1978d) discovered that Rao et al. (1976) had miswritten in their computer program an equation for their kinship category involving the correlation between parental IQ and the offspring's SES, causing them later to make slight corrections in a large number of their original parameter estimates (Rao et al., 1978).

Much of the remainder of this chapter pertains to the second kind of specification error. For the most part these errors consist of omitting quantities that should be included in certain equations. For example, noncorrelation of residual variables is always assumed, yet in many cases quantities representing their correlation could justifiably have been included; quantities representing the effect of an individual's IQ upon his own environment should have been included; and different (rather than equal) values for gene-environment correlation across kinships might have been employed. Furthermore, as will be seen later, different parameters for different intrapair (not residual) environmental similarities greatly affect the resulting h^2 estimates. Finally, the Jencks et al. path model for separated identical twins could have included a fair number of excluded specifications, such as quantities representing effects of social environment on intrauterine environment, or direct sibling effects (see Chapter 3).

12. *That there are no arithmetic errors.* Arithmetic errors in the calculations of Jencks et al., Jensen, and Munsinger (1977, after Kamin, 1978) have already been noted. Later in this chapter I mention what look like arithmetic errors in an important article by Scarr and Weinberg (1977), and in Chapter 5 I point out additional errors on Jensen's part.

13. *That one's data have not been contrived or fabricated.* The review of Burt's data suggests the necessity of stating this requirement (see Chapter 2).

14. *That primary data sources are correctly cited.* Jensen's *elaborate* misquotations from the separated twin studies (see Chapters 2 and 3), Erlenmeyer-Kimling and Jarvik's (1963) errors in compilation (Kamin, 1974), and Herrnstein's and Eysenck's misquotations and incorrect citations (see Chapter 1, note 1) have already been discussed. The interested reader might consult Goldberger (1976a) for an elaborate account of Jensen's misquotations and misstatements from the Burks (1928) study of adopted children and Goldberger (1976b) for an excellent account of Herrnstein's (1973) misquotations from Burks. Additional examples of misquotation will be given later.

I turn now to the role played by the more specific kinds of assumptions in the analysis of the heritability of IQ score, using the technique of path analysis. Path analysis is especially useful in making explicit assumptions that are only implicit in the work of many contributors to the IQ heritability literature.

A LOOK AT PATH ANALYSIS

The most central and fundamental goal of the analysis of the genetic heritability of human intelligence, from Galton through Jensen, has been to investigate whether, assuming all else constant, there is some causal connection between G, the unseen, unmeasured set of genes "for" intelligence, on the one hand, and actual measured IQ score, on the other. At least two closely related strategies for doing so have appeared in the heritability literature. One is the analysis-of-variance strategy borrowed by Jensen from work in biometrical genetics, exemplified by Jensen's basic variance equations (such as equations [2.1] and [2.2]. Another is path analysis. In this section I express the Jensen-like variance equations in the language of path analysis, thereby making it possible to see just how much must necessarily be assumed or guessed at to estimate the heritability of IQ. Given the potential policy implications of IQ heritability analysis, it is not sufficient to justify heritability estimation by means of arbitrary assumptions on the grounds that no heritability analyses could be done without such assumptions. The alternative is to refrain from trying to estimate IQ heritability at all.

The invention of path analysis as a unique tool of multivariate analysis and causal inference is attributed to the pioneering work of the population geneticist Sewell Wright (1921; 1925; 1934). The technique was relatively infrequently used in the behavioral sciences until the mid-sixties, when two seminal articles in sociology by Boudon (1965) and by Duncan (1966) led to its wider adoption. Path analysis has become increasingly popular in psychology (Werts and Linn, 1970), sociology (Blalock, 1971; Heise, 1969; Land, 1969), and the behavioral sciences generally (e.g., Kerlinger and Pedhazur, 1973). The use of path models and causal models in the social and behavioral sciences has been greatly stimulated by the work of Blalock (1962, 1964, 1968, 1971), Goldberger (1972; Goldberger and Duncan, 1973), and others. Recent applications of path analysis to the estimation of IQ heritability are exemplified by the work of Jencks et al. (1972) and the Honolulu group.

The simple path diagram in Figure 4.1, Model 1 is in effect what

Model 1

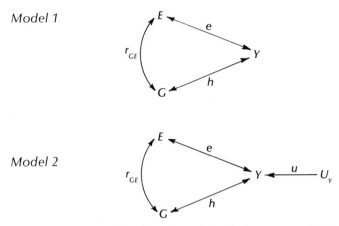

Model 2

Figure 4.1. Simple path models of the sources of IQ variance.

Jencks et al. use as a general guide for their heritability analysis (1972, p. 269; cf. Taylor, 1973b). Such a model is also often employed, though implicitly, by Jensen, Herrnstein, and Eysenck. The unmeasured constructs G and E represent "intelligence genotypes" and all conceivable environmental variables, respectively. The variable Y is measured IQ (not "intelligence"). Following convention, the single-headed arrows represent direction of presumed causation, while the double-headed arrows represent correlation only (r_{GE}), with no necessary causal direction implied. It is thus implausibly assumed, as is standard in the heritability literature, that a person's IQ can causally affect neither his G nor his E. This is the recursiveness assumption of path analysis.[4]

The presumed effects of G and E upon Y are standardized partial regression coefficients (sometimes called "Beta weights"), or *path coefficients*. They are defined as follows:

$$h = \beta_{YG.E}; \text{ thus } h^2 = \beta^2_{YG.E} \qquad (4.1)$$

$$e = \beta_{YE.G}; \text{ thus } e^2 = \beta^2_{YE.G} \qquad (4.2)$$

This clearly states, by definition and convention (Jencks et al., 1972, p. 267; Wright, 1934, p. 164) that h is the regression of IQ on genotype *with environment constant,* or the regression of IQ on genotype for individuals with random genes and identical environments. The path coefficient e is thus the regression of IQ on environment with genotype constant, or the regression of IQ on environment for individuals with exactly the same genes but random environments. Squaring these path coefficients (Wright, 1934, p. 164) yields the proportion of Y's variance explained by G with E constant (h^2, or broad heritability) and by E with G constant (e^2, or environmentability).

If this simple model is considered a correct approximation of reality, then G and E completely determine Y. In this case, the construct E in effect becomes the residual, and any individual's IQ score is weighted by the relative sizes of h and e such that

$$Y = hG + eE, \qquad (4.3)$$

which is the linear structural (multiple regression) equation for Model 1. Since constructs G and E are assumed to be in standard form, the Y-intercept is zero. Equation *(4.3)* as stated here is a relatively standard model in quantitative genetic theory (see, for example, McClearn and DeFries, 1973, p. 185).

In general, as with any structural equation, the coefficient of any given independent variable is a partial coefficient, such that all other independent variables in the same structural equation for Y are assumed constant. If the relationships between Y and G or Y and E are to any degree nonlinear, then equation *(4.3)* will not completely determine Y. Furthermore, if the relationships are to any degree nonadditive, then equation *(4.3)* will not completely determine Y. Thus the assumptions that all relationships are both linear and additive have been introduced. To the extent that these assumptions are implausible, h^2 and e^2 (including the correlation of G and E) will not account for all of Y's total variance.

Although h and e are partial coefficients, portions of the IQ heritability literature, including Jensen's work, inadvertently treat the coefficient h as though it were a *zero-order* (or total) coefficient, such that h (and thus h^2) inadvertently becomes the estimated effect of G on Y without a constant E. Jensen, Herrnstein, and Eysenck make no explicit reference to this distinction, although a contrast between partial and zero-order coefficients is made by Burt (for example, Burt and Howard, 1957b, pp. 103–4). What is actually being estimated, then, is not h, but some zero-order coefficient, say h', such that $h' = r_{GY}$. Generally, heritability estimation procedures have inadvertently produced estimates of the zero-order effect (r_{GY}) rather than of h^2.

The heart of path analysis is the path theorem. By means of this highly useful theorem, any zero-order correlation coefficient (such as r_{GY} or r_{EY}) is expressed as a sum of products of all coefficients along any compound paths between the two variables in question, one variable being regarded as endogenous to (dependent upon) another (exogenous) variable. The basic theorem (Wright, 1934; Duncan, 1966; Land, 1969) is

$$r_{Yi} = \beta_{Yi} = \sum_{\substack{j=2 \\ j \neq i}}^{n} \beta_{Yj} r_{ij}. \tag{4.4}$$

In general, one proceeds by reading back from variable Y to variable i, then forward from i, forming products of all coefficients appearing along all traverses (or compound paths) from i through j to Y, and then summing these products for all possible traverses, not intersecting the same variable more than once in a single traverse.

As a useful illustration, I apply the path theorem to Model 1 in Figure 4.1 and express the total correlations r_{GY} and r_{EY} as follows:

$$r_{GY} = h + er_{GE} \tag{4.5}$$

$$r_{EY} = e + hr_{GE}; \tag{4.6}$$

these are the path estimation equations for Model 1.

In general, as noted by Jencks et al., the squared multiple correlation for all X_i on Y may be expressed as follows:

$$R^2_{Y.Xi} = \sum_{i=1}^{n} \beta_{YXi} r_{XiY}. \tag{4.7}$$

Thus, for Model 1,

$$R^2_{Y.GE} = hr_{GY} + er_{EY}.$$

Substituting from equations (4.5) and (4.6),

$$R^2_{Y.GE} = h(h + er_{GE}) + e(e + hr_{GE}),$$

or more simply,

$$R^2_{Y.GE} = h^2 + e^2 + 2her_{GE}. \tag{4.8}$$

Equation (4.8) is a familiar one in IQ heritability analysis. It partitions the explained portion of the total IQ variance into three components: the portions explained by genotype (h^2), by environment (e^2), and by the correlation (covariance) between genotype and environment ($2her_{GE}$). Since additivity is assumed, interaction variance is presumed to be zero. These three components thus correspond to the genetic, environmental, and covariance components of such variance equations as Jensen's equation *(2.2)*.

It is often incorrectly assumed, as by Jencks et al. (1972, p. 269) and on occasions by Jensen (1967, 1969), that G and E completely determine Y in a general population; in other words, that all the variance in IQ is thereby accounted for, such that

$$1 = h^2 + e^2 + 2her_{GE},$$

and thus that $R^2_{Y.GE}$ is assumed to be 1.00. But if any variance in IQ arises from gene-environment interaction, or nonlinearity, or other unspecified sources of variance in Y (such as measurement error that is not removed through the attenuation correction; or even gene-interaction and environment-interaction covariances), then $(h^2 + e^2 + 2her_{GE}) < 1$, necessarily. (Hogarth, 1974, p. 4, and Taylor, 1973b, p. 445 make essentially the same point, as do Morton and Rao, 1978, p. 17.)

Such unspecified sources of IQ variance can be subsumed under a residual variable, U_Y, as in Model 2 in Figure 4.1, which is simply Model 1 with a residual variable included. It is assumed that this residual variable is uncorrelated with G and E, thus introducing another arbitrary assumption. The structural equation for Y now becomes

$$Y = hG + eE + uU_y, \tag{4.9}$$

and all of Y's variance is thus partitioned as follows:

$$1 = h^2 + e^2 + 2her_{GE} + u^2, \tag{4.10}$$

and thus

$$u^2 = 1 - R^2_{Y.GE}.$$

Equation *(4.10)* is directly analogous to equation *(2.2)*. My quarrel with Jensen, however, centers on just what the residual u^2 is intended to subsume (Jensen assumes that it contains only random measurement error, which is then presumably removed from the total IQ variance by the attenuation correction). An additional unknown quantity (u^2), which potentially subsumes interaction, nonlinearity, nonrandom measurement error, and the like, has been introduced into the estimation procedures. Just what such a residual might contain does not appear to have been fully contemplated by either Jensen or Jencks et al.

A GENERAL COLLATERAL KINSHIP MODEL

Thus far, G, E, and Y have been represented as scores on a single individual. The models in Figure 4.1 are thus intended to represent the sources of IQ variance for a general population of unpaired individuals. In the heritability analysis of IQ data, it has become customary to construct models for pairs of individuals, where the phenotypic (IQ) scores of each pair member are separately specified. Such a model appears here as Figure 4.2. The measured variable Y_1 and unmeasured constructs G_1 and E_1 represent the IQ, genotype, and environment for one member of a pair, and Y_2, G_2, and E_2 represent these for the other pair member.

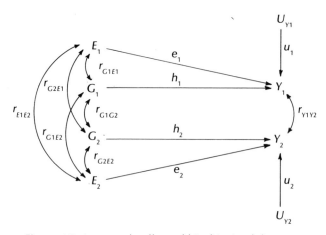

Figure 4.2. A general collateral kinship model.

The model in Figure 4.2 is a general model in that it can be applied to pairs of persons of varying degrees of biological relatedness who are raised either together or apart. The correlation r_{G1G2}, representing the hypothetical correlation of the genotypes of the pair members, will thus vary from 1.00 (for MZ twins) to zero (for completely unrelated individuals). This correlation is in fact the genotypic correlation ρ, the theoretical proportion of shared genes, introduced in Chapter 3 in connection with Jensen's kinship equation.[5] The correlation r_{E1E2}, representing the correlation between the environments of pair members (and thus their intrapair environmental similarity), will theoretically vary from 1.00 (for pairs having absolutely identical environments—highly unlikely even for MZ twins raised together) to zero (for pairs raised absolutely apart in random, uncorrelated environments—

also most unlikely, except of course for unpaired biologically unrelated persons selected at random from a population).

It is often assumed by researchers (especially by Jensen except in Jensen, 1975, 1976) that the environmental correlation r_{E1E2} is equal across kinships (this would be equivalent to setting $r_{E1E2} = 1$ across kinships). The Jencks et al. path model for raised-together kinships in effect does this since their model specifies only one construct (their EF) for our E_1 and E_2 representing common or family environment, thus in effect setting $r_{E1E2} = 1$ across analyzed kinships (1972, p. 296). It is important to note, however, that Jencks et al. assume certain inequalities among certain kinships (MZTs, DZTs, and SIBTs) by allowing for differences in residual (specific or nonfamily) environments U_{y1} and U_{y2}, and thus for differences in the *total* IQ variance across kinships (1972, p. 301). The Jencks et al. model for raised-together kinships also has variables for intrauterine environment (their EU_1 and EU_2), as does their path model for separated identical twins already reviewed (which has one EU). (The Honolulu models of Rao et al., 1974, 1976, 1978, and Rao and Morton, 1978, similarly do not allow for differences in r_{E1E2} across kinships. As will be shown in Chapter 5, they do allow for differences in total IQ variance among some kinships, thus allowing for differences in residual variance.) One kind of assumption, however, is not a substitute for the other.[6]

The correlation r_{Y1Y2}, the kinship correlation itself, is simply the correlation between the IQs of the pair members of a given sample for a given type of kinship. It is the only quantity in the entire model that is actually obtained empirically. In other words, hereditarians such as Jensen are in effect attempting to estimate from kinship data, data on adopted pairs, and data on separated MZs, some h^2 knowing only r_{Y1Y2}. In more technical terms, this kind of model is severely underidentified. The excess of unknown quantities over known quantities (the degrees of freedom) is hopelessly large, which means, generally, that a large range of hypothetical values for the unknowns (including h^2) can be made to fit an observed r_{Y1Y2} or set of such correlations perfectly.

The model in Figure 4.2 assumes zero interaction, recursiveness, that residuals U_{y1} and U_{y2} are uncorrelated with each other or with G_1, G_2, E_1, and E_2 (Jensen, 1976, p. 87, alludes to this last assumption), and that there is no measurement error. Granting these arbitrary assumptions for the sake of argument and keeping in mind that one's results depend upon what one assumes to begin with, the model in Figure 4.2 permits the accomplishment of three tasks:

1. It permits the derivation from the path theorem of the components that underlie the various kinds of kinship correlations reviewed in Chapter 2, as well as the components that underlie the separated MZ twin correlation, and the correlation of unrelated (adopted) pairs raised together.
2. By so doing, important assumptions of heritability analysis that are not at all clearly noted by Jensen, Eysenck, and Herrnstein, and others can be located.
3. It makes it possible to see just how useless Jensen's much-cited kinship equation (equation [2.9]) really is as an index of the heritability of IQ.

Before applying the path theorem to Figure 4.2, three assumptions highlighted by this model should be mentioned:

1. It is sometimes assumed that the gene-environment correlation for one pair member does not differ from the gene-environment correlation for the other pair member, regardless of kinship; that is, that $r_{G1E1} = r_{G2E2} = r_{GE}$. (Exceptions are Jencks et al. and the Honolulu group, who assume that in the case of adopted-natural pairs, the two correlations can differ. The assumption is nonetheless implicit for all other kinships.) Jensen's initial work (for example, 1967; 1969; others) implicitly assumes that $r_{G1E1} = r_{G2E2} = 0$, although two of Jensen's papers (1975, 1976) allow for $r_{G1E1} = r_{G2E2} > 0$. In no case does Jensen ever allow for $r_{G1E1} \neq r_{G2E2}$.

2. Two correlations not previously encountered are the correlations between one pair member's environment and the other pair member's genotype: r_{G1E2} and r_{G2E1}. Jencks et al. allow each of these correlations to be nonzero and unequal. But Jensen does not distinguish these two kinds of correlations at all (e.g., 1975, 1976). He thus necessarily, though implicitly, assumes them to be equal (if they are nonzero). In his earlier work, both are assumed to be zero. It is thus implicitly assumed that

$$r_{G1E1} = r_{G2E2} = r_{G1E2} = r_{G2E1}.$$

This is a big assumption, which amounts to assuming for two individuals: first, that the correlation of one individual's genotype with the other's environment equals the correlation of the first individual's environment with the other's genotype; second, that the correlation of the first individual's genotype with the other's environment equals the correlation of the first individual's genotype with his own environment (and likewise for the other); third, that the correlation of the first individual's genotype with his own environment equals the correlation of the other's genotype with the other's own environment.

3. The third assumption is made, again implicitly, in virtually all research on IQ heritability. Note from Figure 4.2 that no path coefficient appears between Y_1 and Y_2. This amounts to assuming that

$$\beta_{Y1Y2} = \beta_{Y2Y1} = 0, \qquad\qquad (4.11)$$

which is to say, quite implausibly, that the measured IQ of either pair member has no direct causal effect on the measured IQ of the other pair member; that is, that their IQ correlation is attributable entirely to indirect effects—to the variables specified in the model. But surely, for twins, siblings, or unrelated pairs raised together, the IQ test performance of one can affect the IQ test performance of the other. It certainly seems plausible to assume that pairs raised together do talk with one another about such matters.

Such could be the case for pairs raised separately as well. For pairs raised either together or apart, such an effect can be conceptualized as independent of the effects of E_1 and E_2, or of their correlation r_{E1E2}. One cannot envision this influence going from one Y through some E to the other Y, since given the recursiveness restriction, no direct path may go from any Y to any E. If one imagines any such effect that is independent of whatever is subsumed under constructs E_1 and E_2, then either β_{Y1Y2} or β_{Y2Y1} will be nonzero. If either of these coefficients is assumed to be nonzero, then it must be included in any estimation equation for r_{Y1Y2}, introducing another unknown into the estimation system. If either coefficient (or both) is nonzero, then the model in Figure 4.2 is misspecified.

Jencks et al. (1972, pp. 296, 298) allude to the possibility of such direct causal sibling effects for raised-together kinships (but not for separated kinships, such as MZAs), but they do not incorporate such effects into their model. The effect is discussed at some length in Jencks and Brown (1977). Studies of birth order may well pertain to this issue. A study by Zajonc and Markus (1975) has noted extremely consistent effects of birth order and family size on Raven IQ score: The larger the family or the lower (later) the birth order, the lower the IQ. The investigators suggest that older siblings teach younger siblings, which suggests that one direct effect should exceed the other such that $\beta_{Y1Y2} > \beta_{Y2Y1}$, with subscript 1 referring to the younger sibling. Supportive evidence is also seen in Nichols and Broman who found that adjacent siblings are more similar in IQ than nonadjacent siblings (1973, cited in Loehlin et al., 1975, p. 109). Olneck finds that age spacing might affect IQ similarity such that siblings spaced closely (three years apart or less) are more similar ($r_{SIBT} = .516, n = 155$) than siblings spaced farther apart ($r_{SIBT} = .434, n = 197$), though these differences were not statistically significant (1977, p. 130). (Both these studies show the Schwartz and Schwartz treatment effects—the environmental effects on different pairs of equal genetic similarity.)

Application of Wright's path theorem directly to the model in Figure 4.2 is most instructive. Given the recursiveness restriction, if, say, $\beta_{Y1Y2} \neq 0$, then $\beta_{Y2Y1} = 0$ (and vice versa). The path theorem gives

$$r_{Y1Y2} = \beta_{Y1Y2} + e_1 e_2 r_{E1E2} + h_2 e_1 r_{G2E1} + h_1 e_2 r_{G1E2} + h_1 h_2 r_{G1G2}. \quad (4.12)$$

This equation contains 9 separate unknown quantities on the right-hand side; that is, 9 quantities that must be estimated, or about which assumptions, or out-and-out guesses must be made, in order to account for the IQ similarity (r_{Y1Y2}) of any particular collateral kinship pair.

If residuals U_{Y1} and U_{Y2} can be presumed correlated to any degree (as, for example, through any correlation between pair members' non-family environments—which is certainly plausible), then the quantity ($u_1 u_2 r_{UY1UY2}$) would have to be added to equation (4.12), further increasing the number of unknowns to 12. This quantity becomes zero under the already imposed restriction of noncorrelation of residuals, such that $r_{UY1UY2} = 0$. This is of course by definition zero if U_{Y1} and U_{Y2} are considered to represent only specific (noncommon) environments.

The hereditarian literature disposes quite rapidly, though wholly implicitly, of three of these unknowns. First, since β_{Y1Y2} (and β_{Y2Y1}) is assumed to be zero, it drops out of equation (4.12). Second, in actual research practice, pairs are randomly ordered, and hence there is no meaningful way to distinguish h_1 from h_2. Thus it is assumed that $h_1 = h_2 = h$. Third, the same is assumed for e_1 and e_2; thus $e_1 = e_2 = e$. Occasionally, some researchers (e.g., Scarr, 1969, as interpreted by Kamin, 1974, p. 166) have speculated that some intrauterine (thus environmental) asymmetry mechanism may operate in the case of MZ twins, causing certain phenotypes such as birth weight, or even IQ, to develop differently in each twin. For example, one twin may "rob" the other of its blood supply by means of an intraplacental "transfusion," thus causing the birth weight of one twin and thus its IQ to be less than that of the other. As Kamin (1974) notes, however, the evidence for any connection between birth weight and IQ is unconvincing. In fact, Munsinger's (1977) claim of a birth-weight–IQ connection, cited in Chapter 3, not only miscalculates Jensen's difference correlation, but (see Kamin, 1978) depends on Munsinger having guessed the birth weight for a number of twins; arbitrarily assigning IQ scores to Shields's (1962) twins via an unspecified transformation; and simply miscopying some of his data.

Should one wish nonetheless to persist in speculation about intrauterine asymmetries, then one should consider asymmetry phenomena in the social environment as well. There may be complementary role relationships, where for instance one MZ twin becomes a leader, the other a follower (Koch, 1966, alludes to such phenomena). All this implies that $e_1 \neq e_2$ in the case of MZ twins. If such asymmetric predispositions are thought to be themselves heritable (Eysenck, 1967, sees certain personality predispositions as highly heritable), then even $h_1 \neq h_2$ is implied in the case of MZ twins. All this shows that the assumptions that $e_1 = e_2$ and $h_1 = h_2$ are arbitrary.

After making these assumptions, the equation is somewhat simpler:

$$r_{Y1Y2} = e^2 r_{E1E2} + h e r_{G2E1} + h e r_{G1E2} + h^2 r_{G1G2}, \qquad (4.13)$$

showing 6 unknown quantities, fewer than equation (4.12). These are further reduced in number by assuming that $r_{G2E1} = r_{G1E2} = r_{GE}$; and by treating r_{G1G2} as a directly estimable quantity, thus ending up with

$$r_{Y1Y2} = e^2 r_{E1E2} + h^2 r_{G1G2} + 2 h e r_{GE}, \qquad (4.14)$$

an equation fundamental to the contemporary heritability literature. It contains considerably fewer unknowns (h^2, e^2, r_{E1E2}, r_{G1G2}, and r_{GE}) than equation (4.12).

If it is further assumed that $u_1 = u_2 = u$, then one may account for all of Y's variance as follows:

$$
\begin{aligned}
1 &= e^2 r_{E1E2} + h^2 r_{G1G2} + 2 h e r_{GE} + u^2 \\
  &= r_{Y1Y2} + u^2 \qquad (4.15)
\end{aligned}
$$

and thus,

$$u^2 = 1 - r_{Y1Y2}.$$

The assessment of u^2 would employ the IQ correlation for MZ twins raised together such that $u^2 = 1 - r_{MZT}$. In this case u^2 contains non-common environmental effects.[7] More important, since r_{MZT} is presumed to be quite high, at least some variance will always be in a residual, u^2.

Applying equations (4.14) and (4.15) to unpaired individuals from a general population, one can of course assume that $r_{G1G2} = 1$. One can also assume that $r_{E1E2} = 1$, corresponding to the notion of the correlation of an individual's environment with itself. Thus

$$r_{Y1Y2} = r_{Y1Y1} = e^2 + h^2 + 2her_{GE},\qquad(4.16)$$

in which case r_{Y1Y1} becomes (in some formulations) the correlation of an individual's IQ with itself, or test-retest reliability (r_{tt}).[8] Subsuming all else under u^2,

$$1 = e^2 + h^2 + 2her_{GE} + u^2.\qquad(4.17)$$

Note that equation (4.17) is identical to equation (4.10).

If it is assumed, though most implausibly (as by Jensen, 1967, 1975), that there is *no* residual variance, then

$$1 = e^2 r_{E1E2} + h^2 r_{G1G2} + 2her_{GE}.\qquad(4.18)$$

A rather large number of unknowns in equation (4.12) is thus reduced to a smaller, more manageable number by making arbitrary assumptions, some considerably less plausible than others. The process of making assumptions and out-and-out guesses to estimate unknowns like h^2 and e^2 is certainly not unique to Jensen, but necessary to any empirical analysis. But to repeat, given the policy implications of IQ heritability analysis, especially as it bears upon inferred racial or ethnic genetically based differences in intelligence, the practice of making arbitrary assumptions to accomplish empirical solutions may certainly be questioned. Shockley, for example, has based his genocidal policy in significant part upon the kinds of assumptions discussed here.

Measurement Error

1. There is presently much controversy as to whether IQ scores represent *interval scales*. Jensen (1969) argues that they do, since human IQ scores "behave" like other true interval human properties (height, weight) in that their fit to the theoretical normal distribution is reasonably close. But some (Layzer, 1974, among others) have convincingly argued that IQ is only a system for ranking individuals, and is thus only ordinal. There is no metric or unit of measurement tied in closely with a well-founded conceptualization of intelligence. Its reasonable conformity to the normal distribution has little bearing upon whether or not it is an interval scale, since even known ordinal categories can be shown to approximate normality closely.

2. Inadequate *age standardization* (as well as sex standardization) of presently used IQ tests (reviewed in Chapter 3) produces a systematic, not a random, measurement bias.

3. The *cultural bias* of IQ tests is in question. Do IQ tests measure the same concept (or cognitive dimension) across ethnic groups within a society? Do they have the same predictive validity across ethnic groups, particularly black and white? Those arguing in the affirmative (such as Jensen) would seem hard pressed to find no differences, although some (e.g., Gordon and Rudert, 1979) have recently found that IQ tests measure similar dimensions on white and black samples, and that predictive validity in both groups is also similar. But even Herrnstein (1973) admits the IQ scores depend heavily upon linguistic styles, which differ considerably between black and white populations in the United States, even with a control for socioeconomic status (see Dillard, 1972; Hall and Freedle, 1973; Hall and Guthrie, 1979; Hall and Tirre, 1979; Harrison and Trabasso, 1976; Samuda, 1975a, 1975b). In fact, vocabulary tests deliberately loaded in favor of Afro-American culture not surprisingly show higher scores for blacks than for whites, independent of socioeconomic status (Williams, 1975; Samuda, 1975a). I remind the reader again of test administrator Goddard's belief that on the basis of Stanford-Binet tests, "83 per cent" of the Jews arriving in the United States from non-English speaking countries were "found" to be "feeble-minded" (quoted in Kamin, 1974, p. 16).

4. The *dimensionality* of currently used IQ tests is also at issue: Are there several kinds of thinking only one of which is for the most part captured by IQ tests? Jensen has argued at length (esp. 1969) that current IQ tests consistently measure one principal dimension or factor of cognitive functioning, called "g" for "general intelligence," a somewhat antiquated idea borrowed from Spearman. Several theories, however, convincingly demonstrate the existence of several relatively independent kinds of thinking. For example Guilford's (1968) well-known structure-of-intellect model hypothesizes three independent dimensions or factors: operations (cognition, memory, others), contents (figural perception, verbal content, others), and products (use of classes, perception of interrelationships, transformations, others). Several researchers (among them H. Witkin, J. Kagan, D. McClelland, R. Cohen) have hypothesized two basic dimensions: roughly, the analytic (logical or linear) and the integrative (Gestalt or synthetic). This work is summarized in Appendix A. A third conceptualization is

Jensen's (1969, 1972) own "Level" principle, which postulates two kinds of cognitive abilities: Level I (associative or rote learning) and Level II (abstract problem solving).

All these issues, but particularly the last two, bear on the question of whether models for estimating heritability of any cognitive process can justifiably assume only random measurement error (which would result from test unreliability), in which case the popular correction for attenuation is appropriate; or whether systematic (nonrandom) errors underlie IQ measures, in which case the attenuation correction is not appropriate.

As an illustration recall that variables Y_1 and Y_2 in Figure 4.2 represent the IQ scores of collateral pair members. Now assume that each of these is only an imperfect measure of some unobserved concept (intelligence?), and that the accuracy of this connection is a new path coefficient, b, such that

$$Y_1 \xrightarrow{\quad b_1 \quad} y_1$$
$$Y_2 \xrightarrow{\quad b_2 \quad} y_2$$

The higher the b, the more accurate the measure. Thus y_1 and y_2 are now the actually measured, observed IQ scores; Y_1 and Y_2 are the unobserved concepts; r_{y1y2} is the observed kinship correlation (r_{MZT}, r_{MZA}, r_{DZT}, etc.); and thus r_{Y1Y2} is the "true" but unobservable correlation between the "intelligences" of the pair members. Now assume (quite arbitrarily) that $b_1 = b_2 = b$. From the path theorem we see that

$$r_{y1y2} = b_1 b_2 r_{Y1Y2} = b^2 r_{Y1Y2}. \qquad (4.19)$$

Substituting from equation (4.14)

$$r_{y1y2} = b^2 (e^2 r_{E1E2} + h^2 r_{G1G2} + 2her_{GE}),$$

showing that if $b^2 < 1$, then r_{y1y2} will underestimate the "true" correlation r_{Y1Y2} and thus underestimate the sum of its components. Hence, from equation (4.19), the "true" pair correlation is

$$r_{Y1Y2}(\text{"true"}) = \frac{r_{y1y2}}{b^2} = \frac{\text{observed pair correlation}}{\text{measurement coefficient}}. \qquad (4.20)$$

If the test-retest correlation r_{tt} (not r_{tt}^2) is taken as a measure of b^2, then equation (4.20) is precisely the equation for the correction for attenuation for paired data. The attenuation correction is generally given as follows (Walker and Lev, 1953, p. 300):

$$r_{xy}(\text{corrected}) = \frac{r_{xy}(\text{uncorrected})}{\sqrt{r_{xx}} \; \sqrt{r_{yy}}}.$$

But for our paired data, $r_{xx} = r_{yy} = r_{tt}$ and thus simply r_{xy} (corrected) $=$ $[r_{xy}$ (uncorrected)$/r_{tt}]$, which is equation (4.20). If reliability is assumed to be, say, .90, and the attenuation is not justifiable, the resulting artificial inflation of observed correlation $r_{y_1y_2}$ can be quite marked. For example a Jensen-style raw r_{MZA} of .82 becomes (.82/.90), or .91 when corrected in this way.

Now all this presumes that there are no systematic biases operating on either y_1 or y_2 (that is, on either IQ score). Let us assume, for the sake of argument, the following: There are two relatively independent kinds of IQ scores (such as analytic and integrative). Let them be represented as y_1 and y_1', respectively, for pair member 1; and as y_2 and y_2' for pair member 2. Assume further that some unknown variable, call it W (for systematic bias, such as race, SES, sex, or age), affects only y_1' (and y_2') but not y_1 (or y_2). (An example would be if race affected, say, integrative but not analytic thinking.) Incorporating this formulation into the path model in Figure 4.2, and borrowing a strategy suggested in Blalock (1970), Costner (1969), and Heise (1969b), among others, something like the following would be added to the diagram:

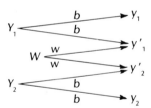

Path coefficient w is the effect of variable W (bias). If $w = 0$, then any correlation between pair members (that is, $r_{y_1y_2'}$, $r_{y_1y_2}$, $r_{y_1'y_2'}$, or $r_{y_1'y_2}$) would be corrected for attenuation as in equation (4.20). (Note also that $r_{y_1y_1'} = b^2 = r_{y_2y_2'}$.) But if $w > 0$ (if some positive systematic bias is present and affects the two kinds of thinking differently), then the correlation $r_{y_1'y_2}$ will be affected, and thus

$$r_{y_1'y_2} = w^2 + b^2 r_{Y_1Y_2},$$
$$= w^2 + b^2(e^2 r_{E_1E_2} + h^2 r_{G_1G_2} + 2her_{GE}),$$

showing that the particular observed correlation $r_{y_1'y_2}$ would be greater than the quantity $(b^2 r_{Y_1Y_2})$ alone, and thus the attenuation correction would not be appropriate. Assuming that w is positive, then direct empirical evidence of such a systematic differential bias would be that the correlation $r_{y'1y'2}$ would be greater than any of the other three empirical correlations taken singly.

To return to equation (4.14), which is so fundamental to the analysis of IQ heritability that one can derive from it three major techniques of h^2 estimation already discussed: the analysis of separated MZ twins; the analysis of adopted pairs; and the various techniques of collateral kinship correlation. I now investigate the extent to which these techniques, each of which necessitates still more assumptions, can be used to estimate the effect of genes upon IQ.

Separated Identical Twins

I begin by assuming recursiveness, linearity, additivity, uncorrelated residuals, random measurement error, that $h_1 = h_2 = h$, and despite any asymmetry mechanisms in the prenatal or postnatal environment, that $e_1 = e_2 = e$. I assume further that the cross-pair gene-environment correlations are equal, such that $r_{G1E2} = r_{G2E1} = r_{GE}$, that $\beta_{Y1Y2} = \beta_{Y2Y1} = 0$, and that $u_1 = u_2 = u$. If residual variables U_{Y1} and U_{Y2} are to any degree correlated, highly plausible if the separation of the MZ twins is not complete, then two additional unknowns ($u^2 r_{UY1UY2}$) must be introduced into the estimation equation. Since the twins are monozygotic, $r_{G1G2} = 1$; thus $h^2 r_{G1G2} = h^2$. From equation (4.14), the observed IQ correlation $r_{Y1Y2(MZA)}$ (that is, r_{MZA}) is

$$r_{Y1Y2(MZA)} = e^2 r_{E1E2(MZA)} + h^2 + 2h e r_{GE(MZA)}. \qquad (4.21)$$

As the correlation or similarity ($r_{E1E2(MZA)}$) between the environments of the presumably separated MZ twins increases, then assuming other components constant, so does the IQ correlation itself. Consequently, any IQ correlation calculated on MZ twins who are incompletely separated will be inflated by the environmental correlation, and artificially so, if, as with Jensen, the observed IQ correlation is treated as a direct estimate of h^2. Any nonzero correlation between the twins' genetic identity and the similarity of their presumably separated environments ($r_{GE(MZA)}$) will also inflate the h^2 estimate if $r_{Y1Y2(MZA)}$ is used to estimate h^2 directly. People who look exactly alike may well get treated similarly in the social environment even if they are separated for a time.

Assume nonetheless, for the sake of argument, that one has a sample of MZ twins who were indeed raised quite separately from the moment of birth, and that one has been able to allocate them to environments completely at random. Only then can one assume, like Jensen, that $r_{E1E2(MZA)} = 0$. Hence

$$r_{Y1Y2(MZA)} = h^2 + 2her_{GE(MZA)}. \tag{4.22}$$

But if the twins' environments are uncorrelated, then $r_{GE(MZA)} = 0$ necessarily, and thus $2her_{GE(MZA)} = 0$. At long last,[9]

$$r_{Y1Y2(MZA)} = h^2. \tag{4.23}$$

This accomplishes Jensen's (1971, 1976) "proof" that: (1) the observed IQ correlation is itself not squared when estimating h^2; and (2) the observed IQ correlation for separated MZ twins is a "direct" estimate of h^2. Jensen was of course not the first to show this; others (for example, Falconer, 1960) have argued that $r_{Y1Y2(MZA)}$ directly measures h^2. It should be noted that certain researchers, for example Spuhler and Lindzey (1967, pp. 403–4, cited in Jensen, 1971, p. 223), have actually squared such observed correlations as an expression of "explained variance." To do so is of course wholly incorrect, as Jensen (1971) himself states, and as is clearly seen from equation (4.23). Finally, note that equation (4.23) has been reached by path analysis rather than by a Jensen-like variance approach.

But Jensen's "proof" is quite meaningless when one considers the extraordinarily long list of arbitrary assumptions that must be made to get the observed IQ correlation to equal h^2. The most demonstrably implausible among these assumptions is the one that $r_{E1E2(MZA)} \cong 0$ (see Chapter 3). Some researchers, though, continue to insist that this environmental correlation is zero for separated MZ twins, and that their observed IQ correlation is a direct estimate of IQ heritability (e.g., Loehlin et al., 1975, p. 287). Even Rao et al. (1976) and Rao and Morton (1978) assume that $r_{E1E2(MZA)} = 0$, as I show in Chapter 5. Note, finally, that if any one of the quantities assumed to be zero ($r_{E1E2(MZA)}$; β_{Y1Y2}; β_{Y2Y1}; the residual r_{UY1UY2}; or any of the four r_{GE} correlations) is in fact greater than zero, then $h^2 < r_{Y1Y2(MZA)}$, necessarily. So much for the correlation between the IQs of separated MZ twins as a "direct" measure of the genetic heritability of IQ.

Adopted Children: Unrelated Pairs Raised Together

The study of separated MZ twins is intended to approximate an experimental condition involving pairs of individuals with identical genes and random environments. The converse case, the study of pairs of individuals with random genes but correlated environments, involves the study of biologically unrelated pairs of individuals raised together in the same family (UNTs), leading to a consideration of $r_{Y1Y2(UNT)}$, the empirically obtained IQ correlation for unrelated pairs raised together. If the pairs can be safely assumed to be biologically unrelated, $r_{G1G2} = 0$ and thus $h^2 r_{G1G2} = 0$, causing h^2 to drop out of the equation. Granting all assumptions leading to equation *(4.14)*, then

$$r_{Y1Y2(UNT)} = e^2 r_{E1E2(UNT)} + 2her_{GE(UNT)}. \tag{4.24}$$

Through selective placement in homes, the *genotypes* of unrelated pairs could conceivably be slightly correlated, in which case $r_{G1G2} > 0$ and $h^2 r_{G1G2}$ would have to reenter equation *(4.24)*. This of course would increase the indeterminacy.

A plausible modification introduced by Jencks et al. and others (e.g., Rao et al., 1976) is that r_{GE} for adopted-natural pairs (UNT-AN) could well differ from r_{GE} for adopted-adopted pairs (UNT-AA). One thus might write

$$r_{Y1Y2(UNT-AN)} = e^2 r_{E1E2(UNT)} + her_{GE(UNT-AN)};$$

and

$$r_{Y1Y2(UNT-AA)} = e^2 r_{E1E2(UNT)} + her_{GE(UNT-AA)} \tag{4.25}$$

But as some (for example, Layzer, 1974, p. 1261; Rao et al., 1976; Rao and Morton, 1978) have noted, if one is interested only in completely unrelated pairs where both are adopted, the dependence of environmental similarity upon their zero genetic similarity is approximately zero; thus $r_{GE(UNT-AA)} \cong 0$, and

$$r_{Y1Y2(UNT-AA)} = e^2 r_{E1E2(UNT-AA)}. \tag{4.26}$$

The environmental correlation (r_{E1E2}) in this case can reflect only in-trapair environmental similarity that arises from being raised in the same family. It can reflect no environmental similarity arising from genetic similarity (that is, from gene-environment correlation). But if one is willing to regard the environmental correlation $r_{E1E2(UNT-AA)}$ as somehow estimable, or, if one is willing to suggest alternative values for this quantity, then the only remaining unknown (e^2) can indeed be found empirically directly from equation *(4.26)*.

What this implies is that environmentability seems more readily estimable from the study of unrelated adopted children than is heritability. There are fewer unknown quantities in the estimation equations for unrelated persons raised together than for MZs, DZs, and sibs raised together, or even for MZs raised apart. This has been noted before, as by Layzer (1974), who uses variance equations. While the quantities r_{G1G2}, all correlations of the form r_{GE}, and h^2 itself are absent from equation *(4.26)*, these quantities must appear in the equations for other kinds of kinships, including separated MZ twins.

One inference from this would be that Jensen, Eysenck, Herrnstein, and others might more productively concern themselves with the estimation of e^2 than with the estimation of h^2. Estimating e^2 from the study of adopted children involves fewer unknown quantities than estimation of h^2. Yet even then, there are difficulties. As noted in Chapter 2, studies of adopted pairs raised together are rare, and the IQ correlation for such pairs has been reported as low as .06 for adopted-natural pairs (Leahy, 1935) and as high as .65 for adopted-adopted pairs (Skodak, 1950). This creates h^2 and e^2 estimates that are difficult to reconcile with one another. I agree with Jencks et al. that a large, well planned study of biologically unrelated pairs randomly placed in the same home is badly needed (1972, p. 367). But even if such an ideal study is conducted, the problem of estimating the environmental correlation and of comparing this correlation to that of other kinships, still remains. In addition, there is the problem of restriction of the total IQ variance for adopted pairs, thus raising problems of whether or not they can be treated as representative of any more general population. One thus wonders just how far such a study would take us toward resolution of the nature-nurture controversy.

KINSHIP CORRELATIONS AND BOGUS HERITABILITY

Assume, again for the sake of argument, recursiveness, linearity, additivity, uncorrelated residuals, random measurement error, and all other assumptions leading to equation *(4.14)*. Using this equation one can write separate equations for different types of kinship pairs. I consider here the method of comparing MZ twins raised together to DZ twins raised together, as done by Jensen (1967, 1969, 1970b, 1972b, 1973, 1975, 1976), as well as comparisons involving siblings and unrelated persons, as done by Jencks et al. using equation *(2.9)*. I also include equation *(4.16)* for unpaired individuals, that is, for the general population. There are thus five equations: for the general population (POP); monozygotic twins raised together; dizygotic twins raised together; full siblings raised together; and unrelated persons raised together (UNT):

$$r_{Y1Y2(POP)} = e^2(1) \qquad + h^2(1) \qquad + 2her_{GE(POP)} \qquad (4.27)$$

$$r_{Y1Y2(MZT)} = e^2r_{E1E2(MZT)} + h^2(1) \qquad + 2her_{GE(MZT)} \qquad (4.28)$$

$$r_{Y1Y2(DZT)} = e^2r_{E1E2(DZT)} + h^2r_{G1G2(DZ)} + 2her_{GE(DZT)} \qquad (4.29)$$

$$r_{Y1Y2(SIBT)} = e^2r_{E1E2(SIBT)} + h^2r_{G1G2(SIB)} + 2her_{GE(SIBT)} \qquad (4.30)$$

$$r_{Y1Y2(UNT)} = e^2r_{E1E2(UNT)} + 0 \qquad + 2her_{GE(UNT)} \qquad (4.31)$$

Since DZ twins and full siblings are equally similar genetically, $r_{G1G2(DZ)} = r_{G1G2(SIB)}$ in equations (4.29) and (4.30). In equation (4.31), if neither pair member is with the natural parents (UNT-AA), then $r_{G1G2(UNT)} = r_{GE} = 2her_{GE(UNT)} = 0$. If the pair contains one adopted and one natural member (UNT-AN), then $r_{GE(UNT)}$ could well be nonzero. Additional equations could be written for other kinds of collateral kinship pairs, such as half-sibs, first cousins, and second cousins, although direct line kinships require other kinds of path models (and thus equations, discussed in Chapter 5). If there is any nonzero correlation between residuals U_{Y1} and U_{Y2} for any kinship considered (representing correlation of nonfamily environments), then u^2r_{UY1UY2} would have to be added to each equation.

Equations (4.27) through (4.31) all contain 11 unknowns if one assumes that $r_{G1G2(UNT)} = r_{GE(UNT)} = 0$. They are h^2 (thus h), e^2 (thus e), 4 environmental correlations (r_{E1E2} for monozygotes, dizygotes, sibs, and unrelated pairs), 4 gene-environment correlations (r_{GE} for the general population, monozygotes, dizygotes, and sibs), and the single genetic correlation r_{G1G2} for both dizygotes and sibs. If $r_{G1G2(UNT)} > 0$ and if $r_{GE(UNT)} > 0$, then there are 13 unknowns. Either way, the number of unknowns exceeds the number of equations linking these unknowns, and the equation system is severely underidentified. There are thus no unique values for all unknowns, including h^2 and e^2, that satisfy all equations. The problem of having the number of kinship equations equal the number of unknowns (thus achieving exact identification), or exceed the number of unknowns (thus achieving the advantage of overidentification), is central to modern heritability analysis.

One strategy for getting around underidentification, as already noted, is to set certain environmental correlations as equal across kinships. This is done by Jensen, though implicitly, for MZT-DZT comparisons (in all his work except 1975, 1976); and in effect by Jencks et al. (1972) for their comparisons among MZTs, DZTs, SIBTs, and UNTs, although Jencks et al. allow for differences in residual variance and thus in total variance (they also undertake a detailed analysis of

parent-child correlations). Rao, Morton, and their colleagues also employ this strategy (discussed in Chapter 5). The r_{GE} correlations are frequently set at zero for MZT-DZT comparisons (Jensen, 1967, 1969,
1972, 1973), or, assumed to be nonzero but nonetheless equal (Jensen,
1975, 1976). There is certainly no reason to expect such gene-
environment correlations to be equal across kinships.

Another strategy, involving equations *(4.27)* through *(4.29)* (explored in Chapter 5) is to assign specific alternative values to $r_{E1E2(MZT)}$
and $r_{E1E2(DZT)}$ (as in Jensen, 1975, 1976) or to $r_{E1E2(MZT)}$ and $r_{E1E2(DZT)}$, as well as
to the r_{GE} quantities (Hogarth, 1974), and then solve the system of
equations simultaneously for sets of these alternative values.

Still another strategy is to increase the number of kinships studied
(including of course parent-offspring pairs), and thus the number of
equations, in an attempt to reduce the gap between the number of
unknowns and the number of equations. This last strategy appears to
be that followed by the Honolulu group. Unfortunately, each time a
new kinship type (and thus an equation) is introduced, at least three
additional unknowns (r_{E1E2}, r_{G1G2}, and r_{GE}) are thereby introduced. (Morton of the Honolulu group himself has said: "The general rule is that
each type of relationship introduces another equation and another assumption, generally in the direction of overestimating heritability,"
1974, pp. 320–21.)

My purpose here is to take a relatively detailed look at pairwise
comparisons of kinships. For convenience, I designate by subscript
any kinship pair to be compared as (1), and any other kinship pair of
lesser biological relatedness as (2). Thus

$$r_{Y1Y2(1)} = e^2 r_{E1E2(1)} + h^2 r_{G1G2(1)} + 2her_{GE(1)} \qquad (4.32)$$

$$r_{Y1Y2(2)} = e^2 r_{E1E2(2)} + h^2 r_{G1G2(2)} + 2her_{GE(2)} \qquad (4.33)$$

On the basis of evidence reviewed in Chapter 3, it seems highly
likely that MZ twins raised together are characterized by a greater degree of similarity in their environment than are DZ twins raised together. It is also extremely probable that DZs raised together have
more environmental similarity than siblings raised together. I again remind the reader of the logic of comparing two kinships of equal genetic similarity. Comparing equations *(4.29)* and *(4.30)*, any observed
positive phenotypic difference ($r_{Y1Y2(DZT)} - r_{Y1Y2(SIBT)}$) can be attributed to
the difference ($r_{E1E2(DZT)} - r_{E1E2(SIBT)}$) if one assumes that $r_{GE(DZT)} = r_{GE(SIBT)}$,
since by definition $r_{G1G2(DZ)} = r_{G1G2(SIB)}$. DZTs and SIBTs do indeed differ in
their respective IQ correlations, a difference Jencks et al. estimate to
be around .10. This observation has interesting implications when one

compares equations *(4.29)* and *(4.30)*: It can be shown that small differences between environmental correlations $r_{E1E2(DZT)}$ and $r_{E1E2(SIBT)}$ imply high e^2 values.[10]

As noted, Burt, Jensen, and Erlenmeyer-Kimling and Jarvik (upon whom Jencks et al. seem to rely in part) have assembled their data so as to minimize the difference in IQ correlation between DZTs and SIBTs. Siblings raised together are probably characterized by a greater environmental similarity than unrelated pairs raised together (Kamin, 1974, reviews evidence of this). If one assumes, quite plausibly, that the environment of a single individual is more "similar" (more correlated with itself) than the environment common to MZ twins raised together, then for equations *(4.27)* through *(4.31)*, it seems likely that

$$r_{E1E1(POP)} > r_{E1E2(MZT)} > . r_{E1E2(DZT)} > r_{E1E2(SIBT)} > r_{E1E2(UNT)}.$$

In general, therefore, it seems likely that for any pair of compared kinships,

$$r_{E1E2(1)} > r_{E1E2(2)}.$$

Recall that Jensen's equation *(2.9)* is

$$h^2 = \frac{r_{AB} - r_{CD}}{\rho_{AB} - \rho_{CD}}.$$

Substituting our symbols for Jensen's, the empirically obtained IQ correlations r_{AB} and r_{CD} become $r_{Y1Y2(1)}$ and $r_{Y1Y2(2)}$, and the unobserved genetic correlations ρ_{AB} and ρ_{CD} become $r_{G1G2(1)}$ and $r_{G1G2(2)}$. Jensen's equation may therefore be written as

$$h^2 = \frac{r_{Y1Y1(1)} - r_{Y1Y2(2)}}{r_{G1G2(1)} - r_{G1G2(2)}} = \frac{\Delta r_{Y1Y2}}{\Delta r_{G1G2}}. \tag{4.34}$$

According to this equation, then, the heritability of a phenotypic variable such as IQ is estimated by taking the ratio of the difference between the observed correlations to the difference between the presumed genetic correlations. It is presumed that $\Delta r_{Y1Y2} \geq 0$. If $\Delta r_{Y1Y2} < 0$, then the heritability estimate will be negative, thus nonsensical, as happened in Chapter 2. It is further presumed that $\Delta r_{Y1Y2} \leq \Delta r_{G1G2}$, for if $\Delta r_{Y1Y2} > \Delta r_{G1G2}$, then the heritability estimate will exceed 1.00, and again be nonsensical, which also happened in Chapter 2.

But Jensen's equation is not a measure of heritability at all. To find out what it does measure, subtract equation *(4.33)* from equation *(4.32)*, to ascertain what the numerator in Jensen's equation (Δr_{Y1Y2}) contains:

$$(r_{Y1Y2(1)} - r_{Y1Y2(2)}) = e^2(r_{E1E2(1)} - r_{E1E2(2)})$$
$$+ h^2(r_{G1G2(1)} - r_{G1G2(2)})$$
$$+ 2he(r_{GE(1)} - r_{GE(2)}),$$

or more simply,

$$\Delta r_{Y1Y2} = e^2 \Delta r_{E1E2} + h^2 \Delta r_{G1G2} + 2he\Delta r_{GE}. \qquad (4.35)$$

Solving to place $(\Delta r_{Y1Y2}/\Delta r_{G1G2})$ on the left-hand side, we obtain:

$$\frac{\Delta r_{Y1Y2}}{\Delta r_{G1G2}} = h^2 + \frac{e^2 \Delta r_{E1E2}}{\Delta r_{G1G2}} + \frac{2he\Delta r_{GE}}{\Delta r_{G1G2}}. \qquad (4.36)$$

This shows that the quantity $(\Delta r_{Y1Y2}/\Delta r_{G1G2})$ is not an index of IQ heritability at all, but instead, an ill-defined index of phenotypic similarity that confounds within it both any excess environmental similarity of kinship (1) over kinship (2) and any excess gene-environment correlation of kinship (1) over kinship (2). If indeed $\Delta r_{E1E2} > 0$ or if $\Delta r_{GE} > 0$, then $(\Delta r_{Y1Y2}/\Delta r_{G1G2})$ yields not heritability, but an artificially inflated index, which I call *bogus heritability*. This bogus quantity is the arithmetic sum of heritability plus two quantities representing artificial inflation resulting from excess environmental similarity and excess gene-environment correlation. Yet some investigators other than Jensen (for example, Loehlin et al., 1975, p. 287) continue to insist that this bogus index measures heritability. (Given the highly implausible possibility that $\Delta r_{E1E2} < 0$, that is that environmental similarity for dizygotic twins exceeds that for monozygotic twins; that environmental similarity for siblings exceeds that for monozygotic twins or even dizygotic twins; and so on, or if $\Delta r_{GE} < 0$, then the quantity $[\Delta r_{Y1Y2}/\Delta r_{G1G2}]$ can actually underestimate heritability.)

Assuming, implausibly, that any two kinships differ in neither their respective environmental correlations nor their respective gene-environment correlations, such that $\Delta r_{E1E2} = \Delta r_{GE} = 0$, then from equation *(4.36)*, $(\Delta r_{Y1Y2}/\Delta r_{G1G2}) = h^2$, thus demonstrating that Jensen's equation requires both these arbitrary assumptions.

Since Jensen's kinship equation will give an artificially inflated estimate of IQ heritability if either, or both, of these assumptions does not hold, it is important to ask how much inflation is involved. How sensitive are h^2 and the difference between bogus heritability and h^2 to these two assumptions? Phrased another way, how much distortion is introduced when $(\Delta r_{Y1Y2}/\Delta r_{G1G2})$ is used to estimate h^2?

Inflation caused by Δr_{GE} itself depends upon the values of h^2 and e^2, since h and e are part of $(2he\Delta r_{GE}/\Delta r_{G1G2})$. To determine the effect of

Δr_{E1E2}, set $r_{GE(1)} = r_{GE(2)} = 0$. Thus $\Delta r_{GE} = 2he\Delta r_{GE} = 0$ in equation *(4.36)*, but this restriction would be only for convenience, since the amount of inflation of bogus heritability resulting from excess gene-environment correlation could be large.

I note, for example, that in comparing, say, MZTs (or even unpaired single individuals) to unrelated adopted-adopted pairs (UNT-AA), since $r_{GE(UNT-AA)} = 0$, then, most interestingly, $\Delta r_{GE} = r_{GE(MZT)}$, thus $2he\Delta r_{GE} = 2her_{GE(MZT)}$, and bogus heritability is thus inflated by the whole of MZT gene-environment covariance. This could go quite high. Using, for example, Jencks et al.'s values of .45 for h^2, .35 for e^2, and .20 for $2her_{GE}$ ($=he\Delta r_{GE}$), then in comparing MZTs (or single unpaired individuals) to UNT-AA, $\Delta r_{G1G2} = 1$, so that inflation is (.20/1) or .20. Comparing DZTs or SIBs to UNTs gives (.20/.50) or .40 as the amount of inflation. Thus inflation of bogus heritability attributable to gene-environment covariance alone could well be very marked.

If for convenience one sets $r_{GE(1)} = r_{GE(2)} = 0$ and assumes that for the general population the residual $u^2 = 0$, then, necessarily, all of the total variance in IQ is due to h^2 or to e^2, and thus $h^2 + e^2 = 1$. Hence $e^2 = 1 - h^2$, and we can solve for h^2 in equation *(4.35)* (and thus in equation [4.36]) as follows:

$$\begin{aligned} \Delta r_{Y1Y2} &= e^2 \Delta r_{E1E2} + h^2 \Delta r_{G1G2} + 0 \\ &= (1 - h^2)\Delta r_{E1E2} + h^2 \Delta r_{G1G2} \\ &= 1\Delta r_{E1E2} - h^2 \Delta r_{E1E2} + h^2 \Delta r_{G1G2} \\ &= h^2(\Delta r_{G1G2} - \Delta r_{E1E2}) + \Delta r_{E1E2}, \end{aligned}$$

and thus,

$$h^2(\Delta r_{G1G2} - \Delta r_{E1E2}) = \Delta r_{Y1Y2} - \Delta r_{E1E2}.$$

Dividing by $(\Delta r_{G1G2} - \Delta r_{E1E2})$, we obtain:[11]

$$h^2 = \frac{\Delta r_{Y1Y2} - \Delta r_{E1E2}}{\Delta r_{G1G2} - \Delta r_{E1E2}}. \tag{4.37}$$

Inserting this into equation *(4.36)*,

$$\frac{\Delta r_{Y1Y2}}{\Delta r_{G1G2}} = \frac{\Delta r_{Y1Y2} - \Delta r_{E1E2}}{\Delta r_{G1G2} - \Delta r_{E1E2}} + (1 - h^2)\frac{\Delta r_{E1E2}}{\Delta r_{G1G2}}$$

$$= \underbrace{\frac{\Delta r_{Y1Y2} - \Delta r_{E1E2}}{\Delta r_{G1Y2} - \Delta r_{E1E2}}}_{h^2} + \underbrace{\frac{\Delta r_{Y1Y2}}{\Delta r_{G1G2}} - \frac{\Delta r_{Y1Y2} - \Delta r_{E1E2}}{\Delta r_{G1G2} - \Delta r_{E1E2}}}_{\text{inflation quantity}} \tag{4.38}$$

Designating bogus heritability as h'^2, and the inflation quantity as f, then:[12]

$$h'^2 = h^2 + f.$$

Bogus heritability is thus a simple sum of heritability and an inflation factor attributable to any difference between kinships (1) and (2) in intrapair environmental similarity—under an assumption of zero gene-environment correlation for both of the compared kinships. Both h^2 and f (where f = bogus heritability minus heritability) can now be expressed in terms of hypothetical values of Δr_{E1E2} for values of Δr_{Y1Y2} and Δr_{G1G2}. This is done in Table 4.1 and Figure 4.3.

Table 4.1 gives the value of bogus heritability (h'^2), heritability (h^2), and artificial inflation (f) for combinations of values of Δr_{E1E2}, Δr_{Y1Y2}, and Δr_{G1G2}. Figure 4.3 plots the inflation quantity as a proportion of bogus heritability (that is, f/h'^2) as a function of Δr_{E1E2} for separate values of Δr_{Y1Y2} and Δr_{G1G2}. Figure 4.3 thus shows what portion of bogus h'^2 results from excess environmental similarity of kin (1) over kin (2), for alternative values of Δr_{Y1Y2} and Δr_{G1G2}.

The values for Δr_{G1G2} were selected to allow not only MZT-DZT and MZT-SIBT comparisons, but MZT-UNT, DZT-UNT, and SIBT-UNT comparisons as well, in the following way: The genetic correlation r_{G1G1} for MZTs was assumed to be 1.00. For DZTs and SIBTs, r_{G1G2} values of .45, .50, .55, and .60 are used. This allows for values in excess of .50 which, according to genetic theory, would allow for effects of dominance and assortative mating.[13] For UNTs, genetic correlation values of .05 (allowing for some genetic correlation arising from selective placement) and (the more plausible) zero are used. These values correspond roughly to values employed in the literature, as by Jensen (1975, 1976), Jencks et al. (1972), Hogarth (1974), Scarr-Salapatek, (1971, 1974), and Scarr and Weinberg (1977, 1978, others). This results in values of Δr_{G1G2} between (1 − .60) or .40 and (1 − .45) or .55 for comparisons of MZTs to DZTs and to SIBTs; between (.45 − .05) or .40 and (.60 − .00) or .60 for comparisons of DZTs and SIBTs to UNTs; and between (1 − .05) or .95 and (1 − .00) or 1.00 for comparisons of MZTs to UNTs. Thus, Δr_{G1G2} values of .40, .45, .50, .55, .60, .95, and 1.00 are employed in Table 4.1 and Figure 4.3.

Differences in the observed IQ correlations between kinships (Δr_{Y1Y2}) are assumed to run from .70 (as, when comparing the IQ correlation of MZTs to that of UNTs) to zero. Hence, values between zero and .70 (in increments of .10) are employed. Conditions of $\Delta r_{G1G2} < 0$ are eliminated, since kinship (1) is always by definition of greater ge-

Table 4.1. Bogus Heritability (h^2), Heritability (h^2), and Inflation (f) for Alternative Values of Δr_{E1E2}, Δr_{Y1Y2}, and Δr_{G1G2} (decimals omitted)

Δr_{Y1Y2}	Δr_{E1E2}	.40 $h^2\ h^2\ f$	.45 $h^2\ h^2\ f$	.50 $h^2\ h^2\ f$	.55 $h^2\ h^2\ f$	.60 $h^2\ h^2\ f$	.95 $h^2\ h^2\ f$	1.00 $h^2\ h^2\ f$
.0	.0	00 00 00	00 00 00	00 00 00	00 00 00	00 00 00	00 00 00	00 00 00
.1	.0	25 25 00	22 22 00	20 20 00	18 18 00	17 17 00	11 11 00	10 10 00
	.1	25 00 25	22 00 22	20 00 20	18 00 18	17 00 17	11 00 11	10 00 10
.2	.0	50 50 00	44 44 00	40 40 00	36 36 00	33 33 00	21 21 00	20 20 00
	.1	50 33 17	44 29 15	40 25 15	36 22 14	33 20 13	21 12 09	20 11 09
	.2	50 00 50	44 00 44	40 00 40	36 00 36	33 00 33	21 00 21	20 00 20
.3	.0	75 75 00	67 67 00	60 60 00	55 55 00	50 50 00	32 32 00	30 30 00
	.1	75 67 08	67 57 10	60 50 10	55 44 11	50 40 10	32 24 08	30 22 08
	.2	75 50 25	67 40 27	60 33 27	55 29 26	50 25 25	32 13 19	30 13 17
	.3	75 00 75	67 00 67	60 00 60	55 00 55	50 00 50	32 00 32	30 00 30
.4	.0	100 100 00	88 88 00	80 80 00	73 73 00	67 67 00	42 42 00	40 40 00
	.1	100 100 00	88 86 02	80 75 05	73 67 06	67 60 07	42 35 07	40 33 07
	.2	100 100 00	88 80 08	80 67 13	73 57 16	67 50 17	42 27 15	40 25 15
	.3	100 100 00	88 67 21	80 50 30	73 40 33	67 33 34	42 16 26	40 14 26
	.4	100 00 100	88 00 88	80 00 80	73 00 73	67 00 67	42 00 42	40 00 40

Table 4.1. (Continued)

				$\Delta r_{G1G2} =$																	
		.40			.45			.50			.55			.60			.95			1.00		
Δr_{Y1Y2}	Δr_{E1E2}	h'^2	h^2	f	h'^2	h^2	f	h'^2	h^2	f	h'^2	h^2	f	h'^2	h^2	f	h'^2	h^2	f	h'^2	h^2	f
.5	.0							100	100	00	91	91	00	83	83	00	53	53	00	50	50	00
	.1							100	100	00	91	89	02	83	80	03	53	47	06	50	44	06
	.2							100	100	00	91	86	05	83	75	08	53	40	13	50	38	12
	.3							100	100	00	91	80	11	83	67	16	53	31	22	50	29	21
	.4							100	100	00	91	67	24	83	50	33	53	18	35	50	17	33
	.5							100	00	100	91	00	91	83	00	83	53	00	53	50	00	50
.6	.0													100	100	00	63	63	00	60	60	00
	.1													100	100	00	63	59	04	60	56	04
	.2													100	100	00	63	53	10	60	50	10
	.3													100	100	00	63	46	17	60	43	17
	.4													100	100	00	63	36	27	60	33	27
	.5													100	100	00	63	22	41	60	20	40
	.6													100	00	100	63	00	63	60	00	60
.7	.0																74	74	00	70	70	00
	.1																74	71	03	70	67	03
	.2																74	67	07	70	63	07
	.3																74	62	12	70	57	13
	.4																74	55	19	70	50	20
	.5																74	44	30	70	40	30
	.6																74	29	45	70	25	45
	.7																74	00	74	70	00	70

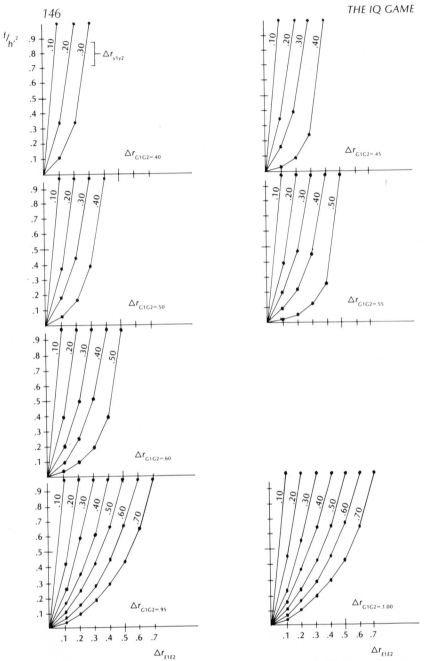

Figure 4.3. Proportion of bogus heritability that is inflation (f/h'^2) as a function of Δr_{E1E2} for alternative values of Δr_{Y1Y2} and Δr_{G1G2}.

netic similarity than kinship (2). Conditions of $\Delta r_{E1E2} < 0$ are also eliminated, since kinship (1) is presumed to be greater than, or equal to kinship (2) in environmental similarity. As a convenience $\Delta r_{Y1Y2} < 0$ is eliminated, since $\Delta r_{Y1Y2} < 0$ gives negative heritability, not by any means an impossibility, as we now know.

For values of Δr_{E1E2},

$$\Delta r_{E1E2} \leq \Delta r_{Y1Y2},$$

which assumes that excess environmental similarity can be no greater than excess phenotypic similarity. (If $\Delta r_{E1E2} > \Delta r_{Y1Y2}$, then h^2 could be negative.) Finally,

$$\Delta r_{Y1Y2} \leq \Delta r_{G1G2},$$

since $\Delta r_{Y1Y2} > \Delta r_{G1G2}$ gives both a bogus heritability and a heritability of greater than 1.00.

Several interesting observations arise from an examination of Table 4.1 and Figure 4.3:

1. Generally, both the size of the inflation quantity (f) and the proportion of bogus heritability attributable to inflation (f/h'^2) increase as Δr_{E1E2} increases. For a given Δr_{E1E2}, inflation increases as Δr_{Y1Y2} decreases. As Δr_{G1G2} increases, the proportion of inflation (for given values of Δr_{E1E2}) also increases. Across all values of Δr_{G1G2}, and for given values of Δr_{Y1Y2}, h^2 declines, and inflation increases, as Δr_{E1E2} approaches Δr_{Y1Y2}. The proportion of inflation becomes especially marked (the curves bend upward) as Δr_{E1E2} nears Δr_{Y1Y2}, such that heritability is zero when two quantities are equal—in which case all of bogus heritability is attributable to inflation by excess environmental similarity. This means that in any instance where the unobserved and unmeasured difference in environmental correlation between two kinships equals their phenotypic (IQ) correlation difference, heritability is zero. Across all values of Δr_{G1G2}, for values of $\Delta r_{Y1Y2} \geq .30$, inflation becomes particularly severe, as $\Delta r_{E1E2} \geq .20$.
2. Generally, if:

$$(\Delta r_{Y1Y2} - \Delta r_{E1E2}) \leq .10,$$

that is, if the excess environmental correlation is within .10 of the excess IQ correlation, the inflation is extraordinarily marked. In fact, an excess environmental correlation approximately equal to the excess phenotypic correlation, for any two kinships is entirely consistent with the hypothesis that the heritability of human IQ is near zero. The possibility that results consistent with zero heritability could arise on the basis of the excess environmental correlation alone (treatment effects for kin of equal genetic similarity) has been investigated by Schwartz and Schwartz (1974, 1975), as noted in Chapter 2.

3. The values of h^2 themselves (Table 4.1), are quite sensitive to even relatively minor differences in Δr_{E1E2}, regardless of the kinships compared. For example, if $\Delta r_{G1G2} = .50$, as with comparisons of monozygotes to dizygotes or sibs, if $\Delta r_{Y1Y2} = .20$, then h^2 goes from .40 to zero as Δr_{E1E2} goes from zero to .20. If $\Delta r_{Y1Y2} = .30$, then h^2 goes from .60 to .33 as Δr_{E1E2} goes from zero to .20. Even if $\Delta r_{Y1Y2} = .40$ (the MZT-DZT difference popularized by Jensen using data from Burt and Erlenmeyer-Kimling and Jarvik), then h^2 goes from .80, Jensen's much-quoted value, to .67 as Δr_{E1E2} goes to .20; to .50 as Δr_{E1E2} goes to .30; and to zero as Δr_{E1E2} goes to .40. This extreme sensitivity of h^2 to only the environmental correlation has been noted by Goldberger (1977) for MZT-DZT comparisons.

The situation is similar if Δr_{G1G2} is .55 or .60, as with comparisons of dizygotes or sibs to unrelated persons. For example, taking Δr_{G1G2} as .60, if $\Delta r_{Y1Y2} = .30$ (which approximates the Jensen, Burt, and Erlenmeyer-Kimling and Jarvik value for such comparisons), then h^2 goes from .50 to .25 as Δr_{E1E2} goes to .20. The sensitivity of h^2 to Δr_{E1E2} persists when $\Delta r_{G1G2} = .95$ or 1.00, as with comparisons of monozygotes to unrelated persons. Taking Δr_{G1G2} as 1.00, even if Δr_{Y1Y2} is .70, h^2 goes from .70 to .57 as Δr_{E1E2} goes to .30; to .50 as Δr_{E1E2} goes to .40; and to .40 as Δr_{E1E2} goes to .50.

Pairwise Comparison of Kinship Categories

The implications of this analysis go beyond the work of Jensen, and pertain to veritably all heritability estimates based on comparisons of MZ twins raised together with DZ twins raised together as well as other pairwise comparisons of kinships. Other indexes besides Jensen's employ the difference in IQ correlation for MZTs and DZTs to estimate heritability. One such index is Falconer's (1960) heritability index, which expressed in our notation is

$$h^2 = 2(r_{Y1Y2(MZT)} - r_{Y1Y2(DZT)}). \qquad (4.39)$$

The use of this index has been advocated by some geneticists (for example, Morton, 1972, p. 257) and by Jensen (1975, 1976). It was used in the Scarr-Salapatek (1971) study, which estimated IQ heritability for black and white MZ and DZ twins, and in Scarr-Salapatek (1974).

Another such index is Nichols's (1965) "heritability ratio":

$$HR = \frac{2(r_{Y1Y2(MZT)} - r_{Y1Y2(DZT)})}{r_{Y1Y2(MZT)}} \qquad (4.40)$$

An older but still-used index is Holzinger's (1929) "heritance" index,

$$H = \frac{r_{Y1Y2(MZT)} - r_{Y1Y2(DZT)}}{1 - r_{Y1Y2(DZT)}} \qquad (4.41)$$

While these three indexes are not monotonically related (as Jensen, 1967, correctly noted), they nevertheless bear certain relationships to each other. In general, all show higher heritability as the difference in the IQ correlations of MZTs and DZTs increases. Furthermore, if $\Delta r_{G1G2} = .5$, then

$$2(r_{Y1Y2(MZT)} - r_{Y1Y2(DZT)}) = (r_{Y1Y2(MZT)} - r_{Y1Y2(DZT)})/.5,$$

in which case Falconer's index is identical to Jensen's. Our consideration of the effects of excess environmental correlation when Δr_{G1G2} is .5 thus applies directly to Falconer's index as well as to Jensen's. As $r_{Y1Y2(MZT)}$ approaches 1.00, the Nichols index approaches Falconer's. Holzinger's index seems ill-conceived even from the outset (Jensen, 1967), although it is identical to Jensen's (and to Falconer's) if $r_{Y1Y2(DZT)} = .5$.

A large number of studies have used comparisons of the IQ correlations of MZ and DZ twins to estimate IQ heritability. Eight studies were listed in Table 2.3, and McClearn and DeFries (1973, p. 207, after Vandenberg, 1971) list seven other studies of white MZ-DZ comparisons, two done in England (Stocks and Karn, 1933, Huntley, 1966), one in the United States (Vandenberg, 1962), two in Sweden (Wictorin, 1952; Husen, 1953), one in France (Zazzo, 1960), and one in Finland (Partanen et al., 1966).

A study by Horn et al. (1976), which compares MZs to DZs on various personality traits, uses Falconer's index. A study by Vogel et al. (1979) in Germany on IQ and various personality measures also uses Falconer's index for MZ-DZ comparisons.

Research by Taubman and his colleagues has investigated the heritability of such "phenotypes" as occupation and earnings. This new area has been dubbed "kinometrics," the study of the determinants of socioeconomic success within and between families (Taubman, 1977). Behrman and Taubman (1976), as well as Behrman et al. (1977) and Taubman (1976), measure several dependent variables on MZTs and DZTs: education, initial occupation, current occupation, and earnings. They permit correlations for the same variable (say, education) for the two members of an MZ or DZ pair, but in addition, they permit correlations between different variables (say, education with initial occupation) for two members of a given MZ or DZ pair. They have concluded that a large component of socioeconomic achievement as measured by these variables is genetically caused. But Goldberger (1978a, pp. 29–32) has shown that their pairwise comparisons of MZTs to DZTs are simply, for any pair of variables, $2(r_{MZT} - r_{DZT})$. He shows furthermore that setting $h^2 = 0$ and allowing variability in a residual term as well as in a parameter for environmental correlation (cf. r_{E1E2}, which the researchers had set $=1$), the fit to their observed correlations is identical to that with $2(r_{MZT} - r_{DZT})$. This shows that h^2 is indeterminate in their analysis. The reader is referred to Goldberger (1977, 1978a) for a complete and lucid treatment. Goldberger (1977, p. 301) has also noted that a paper by Jencks and Brown (1977) in the Taubman (1977) volume defines as "broad heritability" that which comes out to $(\Delta r_{Y1Y2}/\Delta r_{G1G2})$, the Jensen (1967) equation exactly.

These indexes, which are based on pairwise comparisons of kinship categories, have in various forms been lying around in the professional literature and in textbooks for a long time now. They are not indexes of the genetic heritability of human IQ, but ill-defined indexes that are artificially inflated to the degree that the environmental similarity of MZ twins raised together exceeds that of DZ twins raised together, and to the degree that gene-environment correlation for MZs raised together exceeds that of DZs raised together. Precisely the same argument applies for other kinds of pairwise kinship comparisons. To assert that such bogus indexes somehow reflect the presence of unseen genes "for" intelligence in either black or white populations, seems more akin to alchemy or magic than to science.

A large body of research by Scarr and Weinberg (Scarr-Salapatek and Weinberg, 1975; Scarr and Weinberg, 1976, 1977, 1978) is based on the study of families with adopted children (including white families adopting black children), and uses (in Scarr and Weinberg, 1977, 1978) pairwise comparison of categories to estimate the magnitude of IQ heritability. Their research concerns more than attempts to estimate IQ heritability. Thus Scarr and Weinberg (1975; reported also in Scarr

and Weinberg, 1976) studied 101 white families who were above average in SES characteristics (as well as in average IQ) who adopted black
children. They found the mean IQ of the black children to be well
above the population mean of 100. They also found that biological
children of adopting parents tend to score higher in IQ than the
adopted children, a finding they attribute in part to gene-environment
covariance. Their qualifiedly environmentalist conclusion was that
"the social environment plays a dominant role in determining the average IQ level of black children and that both social and genetic variables contribute to individual variation among them" (Scarr and Weinberg, 1976, p. 739).

But while the data collection, testing, and overall design of the
Scarr-Weinberg research are generally impressive, in two of their articles their h^2 estimates rely heavily on simple pairwise comparisons of
correlations among biologically related pairs to correlations among
unrelated pairs. Of particular interest is their study of 101 Minnesota
families containing both biological ($n = 145$) and adopted ($n = 176$)
children, many of whom ($n = 130$) were black. They concluded:
"Comparisons of correlations between related and unrelated siblings
produced negligible heritability values, whereas parent-child data
suggested moderate heritability for the children's IQ differences"
(Scarr and Weinberg, 1977, p. 170). They, however, emphasize the
comparisons of parent-to-natural child correlations to adoptive
parent-to-adoptive child correlations, stating that "on balance, we
concluded that they provide reasonably coherent support for the
moderate heritability of IQ scores in a racially mixed sample of children in Minnesota" (p. 190), and that the h^2 for IQ is in the neighborhood of .40 to .70 (p. 190). They reach similar conclusions in Scarr and
Weinberg (1978).

The central data tables in Scarr and Weinberg (1977) (Tables 4, 5,
and 6 on pp. 179, 180, and 187) all use a single, familiar equation for
estimating h^2, which may be represented as

$$h^2 = 2(r_1 - r_2) = (r_1 - r_2)/.5.$$

Three types of comparisons were made:

1. The IQ correlation between parents and their natural offspring
   (for r_1) was compared to the IQ correlation between parents and
   an adopted child (for r_2). (Correlations were calculated separately for mother, father, and midparent, a *midparent* IQ value

being simply a value halfway between that of both parents; that is, $[IQ_F + IQ_M]/2$.) Note that $r_{G1G2} = .5$ for the first (biologically related) pair, and $r_{G1G2} = 0$ for the second (biologically unrelated) pair, thus giving the above equation. Three coefficients were then calculated, one based on the raw intraclass r's; one correcting the parent's IQ scores for restriction of range in both pair types; and one involving a correction for selective placement (which resulted in lowered values for r_2) as well as a range restriction for r_2 only.

2. The IQ correlation for natural siblings (SIBTs; or r_1) was compared to that for adopted-natural pairs (UNT-AN; or r_2), both with and without a range restriction correction for adopted children's IQ scores.

3. The IQ correlation for natural siblings (again SIBTs for r_1) was compared to that for adopted-adopted pairs (UNT-AA) for r_2, both with and without a range restriction correction.

Similar comparisons are used in Scarr and Weinberg (1978, p. 685), with the exception that $1.6(r_1 - r_2)$ is used to estimate h^2 rather than $2(r_1 - r_2)$, a modification intended by them to adjust for assortative mating. They reach similar conclusions; h^2 comes out to be between .38 and .61 (1978, p. 685, their Table 6, for total WAIS IQ), and they state that "going straight to the heart of the matter, we think that most evidence points to a heritability for IQ of about .4 to .7" (1978, p. 688).

Each comparison ignores any differences between compared categories in environmental correlation r_{E1E2} as well as differences between categories in gene-environment correlation r_{GE}. But gene-environment correlation for biologically related pairs (their r_1) is greater than for unrelated pairs (their r_2)—adoptive parent-adopted child; adopted-natural collateral pairs; and adopted-adopted collateral pairs. This will artificially inflate each heritability estimate with their equation, which they note. But gene-environment correlation is also greater for adopted-natural pairs than for adopted-adopted pairs, and this is ignored by their equation.

Other problems with Scarr and Weinberg (1977) include:

1. Intraclass correlations for 8 pair types in their Tables 4 and 5 are not reported because the number of pairs was less than 20. But this leaves open the possibility that some of them (most were of r_2) were greater than the correlation from which they would be

subtracted (r_1), thus producing negative h^2 values. It is not known whether or not this was the case.

2. There seem to be several arithmetic errors in Tables 4 and 5 (1977, pp. 179–80), which involve the parent-child comparisons. An h^2 value reported as .96 should have been 2(.58 − .07) or 1.02. Another of .38 should be 2(.54 − .33) or .42. There are 6 incorrect h^2 estimates (involving r_1 corrected for range restriction and r_2 corrected for both range restriction and selective placement), which according to the text and table footnotes should be as follows: In their Table 4, a value of .58 should be 2(.49 − .26) or .46; 1.06 should be 2(.54 − .11) or .86; and .72 should be 2(.66 − .33) or .66. In their Table 5, a value shown as .40 should be 2(.48 − .34) or .28; 1.12 should be 2(.48 − .01) or .94; and .66 should be 2(.64 − .34) or .60. Professor Scarr (personal communication, 1979) has indicated that these calculations only correct for "half of" the selective placement effect as described on p. 183 of the article. But if this is so, then 6 *remaining* h^2 calculations involving both types of corrections would be incorrect.

3. In their Table 6, which involves the comparisons of natural sibling pairs to adopted-natural and to adopted-adopted pairs, 3 h^2 values given as "(.00)" are incorrect. They are actually negative, since for each the adopted (or r_2) correlation was greater than the r_1 (biologically related) correlation. One should be 2(.42 − .47) or −.10. Another should be 2(.37 − .49) or −.24. The third should be 2(.37 − .58) or −.42. This last one is moderately high (and absurd), thus directly contradicting the authors' own assertion that the heritability estimates from this same table "were low or negligible" (1977, p. 188).

4. A path diagram on p. 184 shows two exogenous variables: "Ed_{NP}" for education of natural parent of the adopted-away offspring; and "Ed_{AP}" for education of the parent who adopts the particular child. The dependent variable (IQ_{AC}) is the adopted child's IQ score. The correlation between Ed_{NP} and Ed_{AP} is .22 (showing an effect of selective placement); that between Ed_{NP} and IQ_{AC} is .38; and that between Ed_{AP} and IQ_{AC} is .28. One may thus calculate path coefficients (reported by the authors) as follows: For the Ed_{AP}, IQ_{AC} path, [.28 − .38(.22)]/[1 − (.22)^2] or .21 (their path coefficient "e"); and for the Ed_{NP}, IQ_{AC} path, [.38 − .28(.22)]/[1 − (.22)^2] or .34 (their path coefficient "g"). They call coefficient (g) a "genetic path" (p. 184).

But clearly this is *not* a direct path from any genotype to any phenotype; calling it genetic is misleading. It is in fact incorrect to do so: If one assumes (as the authors well might) a correlation (call it c) between natural parent's education, an environmental variable (their Ed_{NP}), and the natural parent's *genotype* (call this G_{NP}), then one can have a direct path (call it d) from G_{NP} to IQ_{AC}. Then by the path theorem,

$$r(Ed_{NP}, IQ_{AC}) = g + cd,$$

or $.38 = .34 + cd$; thus $cd = .04$.

But then the true "genetic path" (d) is indeterminate, unless of course one has a value for gene-environment correlation c, which they do not. Including all paths would give $r(Ed_{NP}, IQ_{AC}) = g + cd + .22e$, only further increasing the indeterminacy of d. Hence, their path g should not be called genetic. As a direct path, it is strictly environmental. While it is interesting that their g exceeds their e (the direct effect of the education of the natural parent upon the IQ of their own adopted-away offspring exceeds the direct effect of the adopting parent's own education), nonetheless a relatively small genetic (d) effect or gene-environment (c) effect is implied.

5. They estimate h^2 in four places in their Tables 4 and 5 by comparing the correlation between Ed_{NP} and IQ_{AC} (a type of correlation for related pair members raised apart, as r_1) to the correlation between the IQ of adopting parent and IQ of adopted child (unrelated pairs raised together, as r_2). But this in effect puts Ed_{NP} in place of what would have to be IQ_{NP} for this particular h^2 calculation. Surely, this presumes a perfect (or near-perfect) correlation between Ed_{NP} and IQ_{NP}, which is not the case. While it is true that IQ data on the natural parents were unavailable, it nonetheless makes little sense to insert Ed_{NP} in place of IQ_{NP} into a grossly oversimplified heritability estimation equation already severely burdened with serious flaws and implausible assumptions.

GENE-ENVIRONMENT INTERACTION

Variables X and Z are said to interact upon variable Y if the slope or form of the relationship between X and Y depends upon the value of Z. In such instances, the relationship between X and Y is said to be

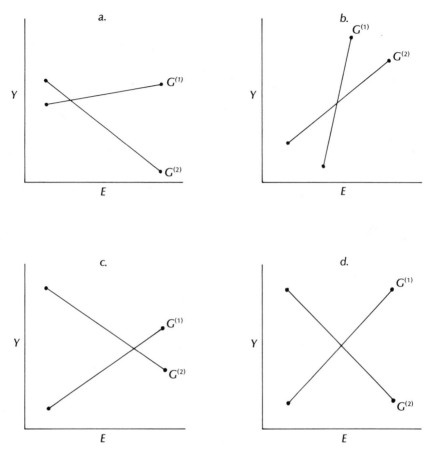

Figure 4.4. Gene-environment interaction. *Source:* Adapted from Lewontin, 1974, p. 405.

conditional upon Z. If, in contrast, the X, Y relationship does not depend upon the value of Z, then additivity exists; or, the X, Y relationship is not conditional upon Z. Gene-environment interaction exists if the slopes of IQ on G are different (that is, nonparallel) for different environments (different "values" of E); or, equivalently, if the slopes of IQ on E are different for different genotypes (different values of G).

Some hypothetical illustrations of gene-environment interaction appear in Figure 4.4, where E is treated as a continuous variable and G as a dichotomous one, thus showing, in each case, two different slopes of Y on E for two different values or "kinds" of G (namely, $G^{(1)}$

and $G^{(2)}$). Geneticists call such different slopes the different *reaction norms* of different values of G upon Y across different environments (see Hirsch, 1972, p. 94; Lewontin, 1974). Such occurrences of different reaction norms happen frequently in studies of animals, where the relationship between some phenotypic trait Y and genetic strain is examined over varying values of some environmental variable, such as temperature. It is evident that there are an infinite number of interaction patterns; Figure 4.4 illustrates four of them.

One cannot of course directly test for the presence of gene-environment interaction in the case of human IQ, since construct G (and E as well) is hypothetical and unmeasured. But if one is willing to argue that some G "for" IQ exists, then one must necessarily consider the problem of additivity versus interaction. Evidence bearing on the additivity-interaction question is admittedly meager. I do not conclude that gene-environment interaction in the case of IQ definitely exists, or even that it is probable, only that it remains a plausible possibility. But there is a Catch-22 to the additivity-interaction issue: To ask whether the relationship between G and IQ is conditional upon environment, one must assume that some G for IQ exists in the first place.

Consider the hypothetical cases illustrated in Figure 4.4. Case *a* shows an overall effect of both genotype and environment upon phenotype Y as well as an effect of their interaction. Thus, all possible effects (two "main" effects and the interaction effect) are illustrated. Individuals characterized by one genotype ($G^{(1)}$) have a higher mean score on Y than individuals characterized by the other genotype ($G^{(2)}$); hence, there is a relationship between genotype and Y. There is also an overall relationship between environment and Y: Generally, Y declines as E increases, since the negative slope of Y on E for $G^{(2)}$ exceeds the smaller positive slope of Y on E for $G^{(1)}$. The environment affects Y in opposite directions for $G^{(1)}$ and $G^{(2)}$, and hence interaction is present.

Case *b* shows an effect of E and of interaction, but no overall effect of G. Individuals characterized by $G^{(1)}$ do not differ, on the average, from individuals characterized by $G^{(2)}$. If the usual analysis-of-variance procedures could be applied in this case, only effects of environment and interaction would be revealed. No effect of G would be found; that is, heritability would be zero. Being a partial coefficient, h^2 reflects the fact that the effects of $G^{(1)}$ and $G^{(2)}$ would be *averaged* across environments. Clearly, such a result would be misleading, since for the lower ranges of E, genotype $G^{(1)}$ is characterized by a lower average score on Y than is genotype $G^{(2)}$, but the reverse is true for the higher ranges of E. This is the interaction.

Case c is the mirror opposite of b, showing effects of genotype ($G^{(2)}$ individuals have higher mean Y scores than $G^{(1)}$ individuals) and also an effect of interaction, but no overall effect of environment. An analysis-of-variance test would show no effect of environment, and the e^2 coefficient would be zero. But such a finding would be misleading, since E indeed affects Y—in opposite directions for $G^{(1)}$ and $G^{(2)}$.

Case d illustrates a particularly intriguing aspect of interaction: Only the interaction effect itself is present. No overall effect of genotype (h^2) or environment (e^2) will show up in an analysis. Such a finding would be misleading, since G affects Y but in equal and opposite directions depending upon E, and E affects Y in equal and opposite directions depending upon G. In principle, it is thus entirely possible to have no main effects from either genotype or environment, but only an effect of their interaction. Thus if interaction is set at zero, as done by Jensen, Herrnstein, Eysenck, Jencks et al., and others, the results can clearly be misleading.

The above illustrations, although hypothetical, are by no means atypical of the kinds of results that experimental geneticists often obtain in studies of animals, many of which reveal strong gene-environment interactions. Jensen has used experimental studies of rats and mice to demonstrate that genetic strain affects their "intelligence" (errors made in going through a maze, e.g., 1969, pp. 30–31). One therefore might be justified in citing evidence he ignores of gene-environment interaction on the intelligence of animals such as mice.

Figure 4.5 presents results from Henderson's (1970) landmark study of six inbred strains of mice, which shows significant interaction. The phenotypic variable (Y) is time (in minutes) that it took the mice to proceed through a maze to food. Genotypes (strains) designated as "A/J" and "RF" differed markedly from those designated "DBA," with genotype DBA requiring the least time. The slopes are far apart but nearly parallel, illustrating additivity with respect to these three genotypes. Note that environment had virtually no effect, especially upon genotypes A/J and RF. But when the other three genotypes are considered ("Balb," "C3H," and "C57BL"), clear interaction is observed. The effect of environment is considerable ("enriched" environments resulted in faster time than "standard" environments), and the overall effect of genotype (for these three genotypes) is small. (Incidentally, if only genotypes A/J, RF, and DBA had been studied, or if only geno-

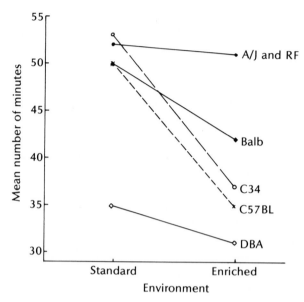

Figure 4.5. An illustration of interaction from a study of mice. *Source:* Adapted from Henderson, 1970, as reported in Erlenmeyer-Kimling, 1972, p. 190.

types Balb, C3H, and C57BL had been studied, then one would have concluded in favor of additivity; interaction would not have been statistically significant.) An obvious caution is suggested by such a study: The relationships between *G* and IQ, and *E* and IQ, must be studied over as full a range of environments and genotypes as possible. Otherwise, one will inadvertently rule out the possibility of finding that additivity exists over a restricted range of *E* (or *G*) and interaction over a fuller range.

The literature contains other studies that similarly show reversed environmental effects for different genetic strains. Erlenmeyer-Kimling (1972) cites several studies of mice that reveal such interaction even for different phenotypic variables: a 1959 study by King and Eleftherious on avoidance behavior as the phenotypic variable; and studies by Newell (in 1967) and by Dixon and DeFries (in 1968) on defecation behavior. Similar interaction results have been found in studies of dogs (Fuller, 1967, 1972, cited in Fuller, 1972) and of fruit flies (Dobzhansky and Spassky, 1944, cited in Lewontin, 1974). Indeed, as Fuller notes, interaction results in animal studies appear to be at least common enough so that "experimental behavior geneticists who look for genotype-environment interactions generally find them" (1972, p. 19).

There exists some evidence for gene-environment interaction on certain human physical traits, such as height. Like IQ, human height is regarded in much of the literature as having high heritability—around .80 or even greater (Dobzhansky, 1973, p. 18). Gottesman and Heston (1972) have shown that Japanese-born women of Japanese-born parents have a mean height (146.1 centimeters) that is less than the mean height (150.5 cm) of United States-born women of Japanese-born parents. In this comparison, parental genotype is in effect assumed constant, and environment is varied—Japan being the "natural" environment and the United States an "enriched" one. Japanese-born men of Japanese-born parents average 158.2 cm in height, while United States-born men of Japanese-born parents average 164.5 cm. That men are generally taller in either environment is evidence of the genetic heritability of height. The difference for women (4.4 cm) and the difference for men (6.3 cm) show a clear environmental main effect. The difference of differences (6.3 versus 4.4) is evidence of gene-environment interaction.

Gottesman (1963) has hypothesized that in the case of human IQ, different intelligence genotypes would show different slopes or different "reaction ranges" across different environments. Jensen used Gottesman's principle to suggest that "higher" intelligence genotypes would reveal a greater reaction range (that is, a steeper slope) than "lower" intelligence genotypes (1969, p. 16, his Figure 11). In this respect, the higher intelligence genotypes would reveal greater potential over a given range of environments. But what Jensen did not clarify is that the Gottesman principle posits gene-environment interaction upon IQ.

Is there any evidence, however indirect, of gene-environment interaction on human IQ scores? Recall from Chapter 2 that studies showing different h^2 values for blacks and whites (for example, Scarr-Salapatek, 1971, 1974) are sometimes interpreted (Dobzhansky, 1973) as suggestive of interaction on the plausible grounds that blacks and whites experience different environments. But recall also the dual character of gene-environment interaction: We do not know whether the pertinent genotype here is the genotype "for" IQ score or the genotype "for" race (say, skin color). (Recall that gene-environment covariance is also subject to this dual character interpretation.)

A much-cited study by Jinks and Fulker (1970) has been quoted, especially by Jensen (1970b, p. 144; 1973, pp. 53–54, pp. 173–74), as supporting a hypothesis of additivity. Jinks and Fulker tested for the presence or absence of gene-environment interaction in human IQ in the following way: Using separated MZ twins, they examined the correlation between the mean IQ of each twin pair (that is, $[IQ_1 + IQ_2]/2$) and the absolute within-pair difference for a pair (that is, $[IQ_1 - IQ_2]$). They argued that the pair mean "reflects" the shared "genotypic value" of a given pair. Clearly, the difference for any pair can result only from environment. Thus, if the IQs of twins with a higher genotypic value (as reflected in their mean IQ score) are more sensitive to environmental differences (that is, have larger IQ differences) than pairs with lower genotypic values, a positive correlation between means and differences should be observed. If, on the other hand, twins with a lower genotypic value are more sensitive to environmental differences, then a negative correlation between means and differences should be observed. In either case, a negative or a positive correlation significantly different from zero would, according to this strategy, suggest interaction, whereas a correlation not significantly different from zero would suggest additivity.

Jinks and Fulker applied their correlation technique to the Shields (1962) sample of 38 separated MZ twins (see Chapter 3, note 2). They found a negative mean-to-difference correlation of $-.28$ for Shields's Mill Hill Vocabulary Test, a correlation of "borderline significance" ($P = .07$). They concluded that some gene-environment interaction for vocabulary ability might exist, but its magnitude would be slight. They obtained an even smaller negative correlation for Shields's Dominoes reasoning test ($-.11$), not significantly different from zero ($P > .10$). Jensen (1970b, p. 144) applied this mean-to-difference correlation to all four of the separated twin studies, thus giving a total N of 122 pairs (see Chapter 3, note 2), and obtained a mean-to-difference correlation of $-.15$, not quite significantly different from zero ($P > .05$). The evidence for gene-environmental interaction, using this technique, thus seems unconvincing. Jinks and Fulker conclude only that the findings encourage "further investigation" (p. 336).

The Jinks and Fulker technique, however, contains several important flaws.

1. The technique assumes a priori the very conclusion it is attempting to demonstrate: that IQ score itself is an indicator of "score" on G. The technique makes sense only if one presumes, as do Jensen, Jinks, and Fulker, that the mean IQ score of a twin pair "reflects their geno-

typic value." This means that high heritability (and zero measurement error) must be presumed beforehand. But heritability is the very thing being estimated. One cannot, as does Jensen, use the very thing being investigated as a foregone conclusion in order to test for something else (presence or absence of interaction) and then in effect conclude that one's initial presumption was therefore correct. (In their otherwise useful discussion of interaction, Plomin et al., 1977, seem to make the same kind of error: They actually use, p. 316, education and other SES variables of the natural midparent of adopted-away offspring as a measure of "genetic factors." Their statement that their technique "is an unusually promising tool," p. 317, seems a bit optimistic.)

2. Even with the presumption that the value of the hypothetical G is somehow "reflected" in the mean score, the mean-to-difference technique tests only for a narrowly defined, simple type of interaction. The technique presumes that higher (or lower) values of G will affect IQ score as a *linear* function of environmental differences. If, for example, some values of G are more sensitive to environment than others, and if such G values are scattered randomly throughout the range of IQ scores (or at least scattered in some nonmonotonic way), this genotype-environment interaction would not be detected by the Jinks-Fulker technique. This difficulty has been noted by Loehlin et al. (1975, pp. 86–87).

3. Jensen states that the mean-to-difference correlation across all four twin studies of $-.15$ "is not significantly different from zero" (1970, p. 144). A correlation of $-.15$ based on an N of 122 pairs is indeed not significant at the .05 level of probability, but it *is* significant at the .10 level ($F_{1,120} = 2.8$, $.05 < P < .10$). This qualifies as at least "borderline significance," using Jinks and Fulker's own criterion for significance. If the .10 criterion had been used, one would have concluded that at least some evidence exists for interaction.

This is not the only place in which Jensen seems to choose significance levels depending upon what he wishes to demonstrate. For example, in the same article, he states that the standard deviations of the twins' absolute differences "are nonsignificant at the .01 level" (1970, p. 141). What he does not say is that the differences are significant at the .02 level. Furthermore, as Schwartz and Schwartz (1976) show, the *means* of the four samples differed significantly, which made pooling of the samples inappropriate.

4. Jensen transformed the raw test scores from the Shields study to have a mean of 100 and a standard deviation of 15 points, but whether

one finds a positive, negative, or near-zero correlation between pair means and differences is dependent upon the scale transformation used. Kamin notes that when using raw scores from Shields's Mill Hill Vocabulary Test, a negative mean-to-difference correlation can be obtained, but transforming the same scores into percentile scores and then into IQ scores can yield a mean-to-difference correlation of zero (1974, pp. 151–52). In the first instance, one would conclude in favor of interaction; in the second, in favor of additivity. Hence, whether or not one obtains "evidence" of interaction may be a statistical artifact.

Nonadditive Path Models: An Illustration

To the extent that gene-environment interaction in the determination of human IQ remains plausible, one might consider the implications for heritability estimation from a model that explicitly includes interaction as a construct. Following a present convention in the construction of path models and structural equations in the behavioral sciences (for example, Althauser, 1971; Blalock, 1965, 1966, 1967a, 1967b, 1968, 1969; Cohen, 1968; Graybill, 1961; Hornung, 1977; Kerlinger and Pedhazur, 1973; Smith and Sasaki, 1979; Southwood, 1978; Taylor, 1973a, 1973b; Taylor and Hornung, 1979), and in behavioral genetics as well (for example, Hogarth, 1974, p. 12; McClearn and DeFries, 1973, p. 185; Plomin et al., 1977) interaction may itself be treated as an unmeasured variable or construct, I, such that by definition,

$$I = f(G,E), \qquad\qquad (4.42)$$

where I is defined as some exact but nonlinear function of constructs G and E. For example, according to one convention, one might define $I = G \times E$, implying that the phenotypic variable is hypothesized to be a multiplicative function of the joint operation of genotype and environment. Or, one might define $I = |G - E|$, suggesting that the phenotypic variable is some function of the absolute difference between genotypic and environmental values. (This means that one cannot always reduce interaction to $\log G + \log E$, Jensen, 1973, p. 213, to the contrary.) Such definitions of I will generally result in nonperfect (<1) but nonzero linear correlations between I and G, and between I and E. I wish to stress, however, that the construct I here represents an "effect" of interaction, and will be only an imperfect measure of conjoint effects of G and E. Indeed, some (for example, Bohrnstedt, 1976; Bohrnstedt and Marwell, 1978; Taylor and Hornung, 1979) have argued that such definitions of interaction (such as a product) will overstate main effects and understate (attenuate) interaction effects. There are virtually a limitless number of ways to define I as a function of G and E, as long as I is not an exact linear function of some combination of values on G and E.

The variable I has its own path coefficient, i, such that

$$i = \beta_{YI.GE},\qquad(4.43)$$

and the structural equation for Y (that is, for measured IQ score) thus becomes

$$Y = hG + eE + iI + uU_Y,\qquad(4.44)$$

suggesting that the effect of interaction (i) on IQ is ascertained while both genotype and environment are assumed constant. Thus, i^2 is the proportion of the total variance in IQ that is explained by interaction. Equation (4.44) expresses a given individual's IQ score as a weighted function of his or her genotype, environment, interaction, and a residual (u).

Treating interaction as a variable results in the realization that the definition of both heritability and environmentability as (squared) path coefficients requires that interaction itself be held constant. This would be true by definition, since with I explicitly introduced as a determining (independent) variable,

$$h = \beta_{YG.EI};\qquad(4.45)$$

and

$$e = \beta_{YE.GI}.\qquad(4.46)$$

A collateral kinship model for coefficients h, e, i, and u is given in Figure 4.6. This model merely adds an interaction construct to the additive path model in Figure 4.2, in effect removing interaction variance from the residual of the model in Figure 4.2. Four new unobserved (thus unknown) correlations are thereby introduced: r_{G1I}, r_{E1I}, r_{G2I}, and r_{E2I}. As before, this model is applied across different kinds of kinship pairs, raised together or apart. A consideration of interaction in this context is particularly instructive, showing what additional unknowns arise from explicit consideration of interaction.

The introduction of an interaction construct in this manner is purely speculative, but the nonadditive model in Figure 4.6 is parsimonious: Only one interaction construct is introduced. This means that interaction is assumed to be of the same type or form regardless of whether it is defined by G_1 and E_1; by G_1 and E_2; by G_2 and E_1; or by G_2 and E_2.

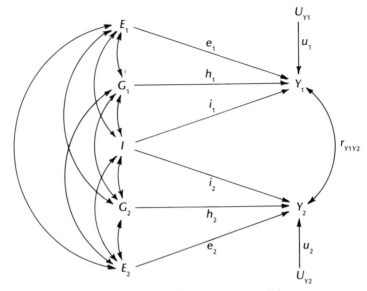

Figure 4.6. A nonadditive kinship model.

Otherwise, at least four interaction variables would have to be specified. If indeed the type of interaction depends upon which G is interacting with which E, then three additional Is and the appropriate correlations would have to be introduced into the system. This would of course only increase the overall number of unknowns. As with the model in Figure 4.2, recursiveness, noncorrelation of residuals U_{Y1} and U_{Y2} with each other or with all five constructs (G_1, G_2, E_1, E_2, and I), that $\beta_{Y1Y2} = \beta_{Y2Y1} = 0$, and that the genetic correlation r_{G1G2} is directly estimable are all assumed.

Given these assumptions, application of the path theorem directly to Figure 4.6 yields

$$
\begin{aligned}
r_{Y1Y2} = \; & e_1 i_2 r_{E1I} + h_2 e_1 r_{G2E1} + e_1 e_2 r_{E1E2} \\
& + h_1 i_2 r_{G1I} + h_1 h_2 r_{G1G2} + h_1 e_2 r_{G1E2} \\
& + i_1 i_2 + h_2 i_1 r_{G2I} + e_2 i_1 r_{E2I}.
\end{aligned} \tag{4.47}
$$

As with Figure 4.2, it is assumed that $h_1 = h_2 = h$ and that $e_1 = e_2 = e$. If the analogous assumption is introduced that $i_1 = i_2 = i$, then equation (4.47) becomes

$$
\begin{aligned}
r_{Y1Y2} = \; & eir_{E1I} + her_{G2E1} + e^2 r_{E1E2} \cdot \\
& + hir_{G1I} + h^2 r_{G1G2} + her_{G1E2} \\
& + i^2 + hir_{G2I} + eir_{E2I}.
\end{aligned} \tag{4.48}
$$

This is the most parsimonious equation for any kinship type unless further assumptions are introduced. One might assume, as does Jensen, that cross-pair gene-environment correlations are equal such that $r_{G1E2} = r_{G2E1} = r_{GE}$. Analogously, one might introduce two heretofore unencountered assumptions, namely

$$r_{E1I} = r_{E2I} = r_{EI};$$

and

$$r_{G1I} = r_{G2I} = r_{GI}. \qquad (4.49)$$

This means assuming that for any pair, the correlation of one pair member's E with I equals the correlation of the other pair member's E with I; and further, that the same holds for the correlation of each pair member's G with I, which may well be an implausible restriction. The equation for r_{Y1Y2} then becomes

$$r_{Y1Y2} = e^2 r_{E1E2} + h^2 r_{G1G2} + i^2 + 2her_{GE} + 2hir_{GI} + 2eir_{EI}. \qquad (4.50)$$

If $u_1 = u_2 = u$, then

$$1 = r_{Y1Y2} + u^2. \qquad (4.51)$$

For the general population of unpaired individuals, it is assumed as before that $r_{G1G2} = r_{E1E2} = 1$. If one assumes further that all of Y's variance is attributable to heritability, environment, interaction, and the covariances in equation (4.50) and thus there are no residual effects resulting from measurement error, nonlinearity, additional interactions, and the like, then

$$1 = e^2 + h^2 + i^2 + 2her_{GE} + 2hir_{GI} + 2eir_{EI}. \qquad (4.52)$$

Thus the total IQ variance has finally been partitioned into those portions explained by genes, environment, gene-environment interaction, and three kinds of covariances—resulting from the correlation between genotype and environment, genotype and interaction, and environment and interaction. Equation (4.50) is a nonadditive version of equation (4.14). In assuming additivity, Jensen, Herrnstein, Eysenck, Jencks et al., and others have implicitly assumed that i^2, as well as r_{E1I}, r_{E2I}, r_{G1I}, and r_{G2I}, are all zero (although Jencks et al. do attempt to justify the additivity assumption).

One might envision separate equations (from equation [4.50]) for separate kinship categories. But this would give rise to the possibility that correlations r_{GI} and r_{EI}, and conceivably even coefficient i, are different across different kinships. If indeed interaction is hypothesized to be nonzero, these three quantities would then enter any equation for pairs of individuals raised separately as well—notably, equations *(4.21)* through *(4.23)* for separated MZ twins. This of course would only further increase the indeterminacy of other unknowns such as h^2 and e^2.

SUMMARY

In this chapter, I have explored the unknowns about which assumptions must be made to arrive at some quantitative estimate of how some hypothetical gene or set of genes might affect the score on an IQ test. The list of assumptions is long, and many of them are wholly arbitrary and often implausible as well.

1. Certain assumptions that are relatively standard to multivariate analysis in the social sciences also apply to the study of IQ heritability. In particular, IQ heritability analysis requires the assumptions of linearity, recursiveness (one-way causation), little or no sampling bias, additivity, noncorrelation of residual variables with each other, noncorrelation of residual variables with main variables, random and near-zero errors of measurement in the dependent variable, and at least a reasonable approximation of "all else constant," especially with regard to the social environment.

2. With the aid of Wright's path theorem, and focusing upon the analysis of kinships taken one at a time, it was possible to state equation *(4.14)*, which is fundamental to heritability analysis— but only after specifying several important assumptions that are often only implicit in the literature: (a) the assumption that gene-environment correlation is the same for both members of a kinship pair (that is, that $r_{G1E1} = r_{G2E2}$); (b) that the correlation between one pair member's genes and the other's environment exactly equals the correlation between the former's environment and the latter's genes (that $r_{G1E2} = r_{G2E1}$); (c) that all four of these gene-environment correlations are equal to each other *and* to the gene-environment correlation characterizing a general population (the r_{GE} for unpaired individuals); (d) that the measured IQ score of one pair member has no direct causal effect on the measured IQ score of the other pair member that is independent of whatever is subsumed under constructs E_1 and E_2 (that $\beta_{Y1Y2} = \beta_{Y2Y1} = 0$); and (e) that neither genetic nor pre- or postna-

tal asymmetries or role complementarities exist in the case of MZ (or DZ) twins, such that both genetic and environmental effects for one pair member equal these for the other pair member (the two assumptions that $h_1 = h_2$ and that $e_1 = e_2$).

3. Focusing on kinships taken two at a time, and using equation (4.14) as a starting point, Jensen's kinship equation was shown to be not an index of IQ heritability, but instead a bogus index artificially inflated by (that is, sensitive to) any excess environmental similarity of kinship 1 over kinship 2 as well as any excess gene-environment correlation of kinship 1 over kinship 2. Focusing upon not only MZT-DZT comparisons, but upon comparisons involving siblings and adopted unrelated persons as well, this artificial inflation as a function of excess environmental similarity (Δr_{E1E2}) for alternative values of excess observed IQ correlation (Δr_{Y1Y2}) and presumed excess genetic similarity (Δr_{G1G2}) was explored. Zero or near-zero heritability is implied as the (unmeasured) excess environmental similarity nears the (measured) excess IQ correlation in magnitude. Such inadequacies are properties of several popular heritability indexes, such as Jensen's, Falconer's, Nichols's, and Holzinger's. Many recent studies employ pairwise comparison of kinships to estimate h^2: Besides the many studies of MZT-DZT differences (including the research on income of the kinometrics group), there is the Scarr and Weinberg research on adopted children. Each time a different kinship (and thus equation) is added to any system of two or more equations, a minimum of three new unknowns (r_{E1E2}, r_{G1G2}, and r_{GE}), and probably more than three, are thereby introduced. Generally, then, in the heritability literature, the number of unknowns greatly exceeds the empirical information necessary to solve for them.

In terms of plausibility, these assumptions can be divided into two categories: those there is strong reason to suspect are quite implausible, often because of direct empirical evidence; and those, less obviously implausible, that nevertheless seem dubious. In the first category are the following:

1. That the intrapair environmental similarity characterizing the actually studied separated MZ twins is zero or approximately zero.

2. That the intrapair environmental similarity characterizing kinship pairs raised together is equal across any two or more kin-

ship types. In particular, I question the assumption that this quantity is equal for MZs raised together and DZs raised together and that it is equal for DZs raised together and siblings raised together. I also question the assumption that the intrapair environmental similarity of MZ, DZ, and sibling pairs raised together in the same family exactly equals that of unrelated pairs of adopted persons raised in the same family. Assumptions about intrapair environmental similarity are conceptually as well as empirically distinct from assumptions about residual variance.

3. That the dependence of environmental similarity upon the genetic identity of separated MZ twins is zero; that is, that $r_{GE} = 0$ for separated MZ twins. The separation of the MZ twins in the studies in question was quite incomplete. If the separation is incomplete, then the MZ twins, who look exactly alike, probably get treated similarly in their imperfectly separated environments. This will produce at least some nonzero dependence of environmental similarity on their genetic similarity, and thus a nonzero r_{GE}.

4. That the dependence of environmental similarity upon genetic similarity (and thus r_{GE}) is equal across two or more types of raised-together kinships. Much of the literature being considered (particularly most, though not all, of Jensen's work) implausibly assumes, either implicitly or explicitly, that this gene-environment correlation is zero (thus equal) across kinships being analyzed. If it is assumed to be nonzero, then it is assumed to be equal across kinships (as in all of Jensen's work, including his most recent). One can be quite confident that $r_{GE} > 0$ for MZTs and that $r_{GE} \cong 0$ for adopted-adopted unrelated pairs. (Any environmental similarity for such pairs cannot arise from their biological similarity, but only from being in the same family.) Thus, r_{GE} for some kinds of kinships (such as MZTs) will exceed r_{GE} for others (such as UNT-AAs). In general, therefore, it is probably implausible to assume that r_{GE} is exactly equal across other kinds of kinships. This is a major flaw in the research of Scarr and Weinberg who compare related pairs (siblings; parent-child pairs) to unrelated pairs (adopted-natural pairs; adopted-adopted pairs; and adoptive parent-child pairs).

5. That one's own IQ score has absolutely no effect whatever upon how one is treated in one's own environment. This is a grossly implausible assumption, which arises from the as-

sumption of recursiveness. The assumption is wholly implicit in all research and models discussed with the exception only of Jencks et al. (1972).

6. That for any kind of kinship considered, the measured IQ of one pair member has no direct causal effect upon the measured IQ of the other pair member.

7. That the standard IQ tests used in the research considered measure the same concepts ("dimensions") on white and middle-class persons as on nonwhite (especially black) and non-middle-class persons.

8. That the only form of measurement error that one need correct for is random error. This is presumably removed by the correction for attenuation (that is, for test-retest unreliability) used extensively by both Jensen and Jencks et al. Certainly, some non-random errors may well enter, especially when one attempts to compare white samples to black samples. In such cases, the attenuation correction is wholly inappropriate.

I am skeptical, but less certain, about the following assumptions:

9. That the effects of "intelligence genotype" (should such exist) and social environment upon IQ score are additive.

10. That the amount of sample bias in kinship studies, studies of separated MZ twins, and studies of adopted pairs, is low or minimal. Since the range of total variance for adopted pairs (including MZAs) is restricted, their representativeness of a *general* population is called into question.

11. That a residual (nonfamily) construct need not be included in any model for a general population or for MZ twins raised together.

12. That residuals U_{y1} and U_{y2} for pair members, for all types of kinships analyzed, both raised-together and raised-apart, are uncorrelated with each other.

13. That residuals U_{y1} and U_{y2}, for all types of kinships analyzed, both raised-together and raised-apart, are uncorrelated with G_1, G_2, E_1, and E_2. Is it really plausible to assume no correlation at all between family environment and nonfamily environment for any given individual or pair of individuals? Is it really plausible to assume no correlation between any environment (which includes nutrition) and intrauterine environment (as do Jencks et al.)?

14. That cross-pair gene-environment correlations are equal for any

given kinship; and further, for any kinship considered, that the gene-environment correlations for each pair member are equal to each other as well as to the gene-environment correlation for a general population of unpaired individuals.

15. That the gene-to-IQ causal path for one member of a pair (h_1) equals that for the other member (h_2), for all kinships considered; and similarly, for the environment-to-IQ causal path (e_1 and e_2).

In sum, I have stressed the indeterminacy of h^2 (and to a lesser extent, e^2), for human IQ. Given the number of required assumptions and the implausibility of the important ones, I see no reason to argue that the genetic heritability of human IQ is a reliably estimable quantity. For the three methods considered—the study of separated MZ twins, unrelated adopted individuals, and kinship correlation considered two at a time—the number of unknowns consistently exceeds the information necessary to solve for crucial unknowns like h^2 and e^2. Making wholly arbitrary and implausible assumptions about these unknowns merely for the purpose of cranking out estimates of IQ heritability cannot be justified on the basis that no estimates could be made without the assumptions. The alternative is not to waste one's time in attempting to estimate the heritability of IQ.

Chapter 5

Studying Kinships Many at a Time

Chapter 4 took a relatively detailed look at procedures for estimating IQ heritability based on comparisons of kinships taken two at a time. In this chapter, I consider the estimation of IQ heritability by means of the analysis of several kinships taken simultaneously. I consider first a three-kin analysis by Jensen and next the research of N. Morton, D. C. Rao, and their colleagues—the Honolulu models, a useful system for the path analysis of several kinships at a time. Because of the clarity and detail of the Honolulu formulation, their assumptions can be easily located. I also undertake an illustrative analysis. Finally I take a brief look at some of the research of what has come to be known as the Birmingham group, contrasting their specifications with those of the Honolulu group.

SOLVING THREE EQUATIONS:
AN EXAMPLE FROM JENSEN

In two articles, Jensen (1975, 1976) has made a brief but flawed attempt to investigate heritability and environmentability via the simultaneous solution of three kinship equations, seeking to explore how h^2, e^2, and r_{GE} behave as a function of alternative values of $r_{E1E2(MZT)}$, $r_{E1E2(DZT)}$, and $r_{G1G2(DZ)}$. But Jensen's calculations are incorrect. I look now at his technique and his calculations.

Jensen states that for the general population (unpaired individuals), $r_{Y1Y2} = .95$ (corresponding to test-retest reliability), and from Erlenmeyer-Kimling and Jarvik, that r_{Y1Y2} for MZTs is .87, and for like-sex DZTs is .56 (1976, pp. 96–98). On the basis of the analysis in Chapter 4, the following three equations, all based on equation *(4.14)*, can be stated:

$$.95 = e^2 \qquad + h^2 \qquad + 2her_{GE(POP)} \qquad (5.1)$$

$$.87 = e^2 r_{E1E2(MZT)} + h^2 \qquad + 2her_{GE(MZT)} \qquad (5.2)$$

$$.56 = e^2 r_{E1E2(DZT)} + h^2 r_{G1G2(DZ)} + 2her_{GE(DZT)} \qquad (5.3)$$

These equations can be written in terms of actual variances and covariances rather than in terms of variance proportions. Recall that

$$h^2 = \frac{\sigma_G^2}{\sigma_Y^2} \text{ (the proportion of } Y\text{'s variance explained by } G) \qquad (5.4)$$

$$e^2 = \frac{\sigma_E^2}{\sigma_Y^2} \text{ (the proportion of } Y\text{'s variance explained by } E) \qquad (5.5)$$

where σ_G^2 is the variance in IQ (not a proportion) explained by G, and σ_E^2 is the variance in IQ explained by E. Following Jensen (1976, p. 98), the standard deviation for IQ in a general population (σ_Y) is 15 points, and the total variance is thus $\sigma_Y^2 = 15^2 = 225$. Assuming a test-retest reliability of .95, equations (5.1) through (5.3) can be expressed in terms of covariances, thus replacing .95 with .95(225) or 213.75 for the general population; .87 with .87(225) or 195.75 for MZTs; and .56 with .56(225) or 126.00 for DZTs. Equations (5.1) through (5.3) can be re-stated as

$$213.75 = \sigma_E^2(1) \qquad + \sigma_G^2(1) \qquad + 2\sigma_G\sigma_E r_{GE(POP)} \qquad (5.6)$$

$$195.75 = \sigma_E^2(r_{E1E2(MZT)}) + \sigma_G^2(1) \qquad + 2\sigma_G\sigma_E r_{GE(MZT)} \qquad (5.7)$$

$$126.00 = \sigma_E^2(r_{E1E2(DZT)}) + \sigma_G^2(r_{G1G2(DZ)}) + 2\sigma_G\sigma_E r_{GE(DZT)} \qquad (5.8)$$

These are Jensen's equations (33)–(35) (1976, pp. 96–98).

Jensen assumes here, as in all of his work, that gene-environment correlation is the same for the general population, MZTs, and DZTs; namely, that $r_{GE(POP)} = r_{GE(MZT)} = r_{GE(DZT)}$, an assumption I have quarreled with before. Once again, it is assumed here for the sake of argument. If one were to hypothesize specific values for $r_{E1E2(MZT)}$, $r_{E1E2(DZT)}$, and $r_{G1G2(DZ)}$, one would have three equations written in terms of three unknowns (σ_E^2, σ_G^2, and r_{GE}), thus exactly identifying the system and enabling empirical solution for these three unknowns. This procedure in principle provides a method for estimating the amount of gene-environment correlation, r_{GE}—if r_{GE} is assumed to be the same across kinship categories. Thus, contrary to some critics of Jensen (for example, Layzer, 1974, p. 1259), and as Jensen (1976, p. 92) himself has correctly pointed out, one need not necessarily assume that r_{GE} is zero across kinship categories to solve for h^2 and e^2 in such equation systems.

Jensen next assigned the following hypothetical values to the following parameters: $r_{E1E2(MZT)} = .90, .80,$ and $.70; r_{E1E2(DZT)} = .90, .80, .70,$ and $.60;$ and $r_{G1G2(DZ)} = .50, .54, .58, .60,$ and $.70.$ This gives $5 \times 4 \times 3$ or 60 possible combinations of values. He then solved equations (5.6)–(5.8) simultaneously for σ_E^2 (thus σ_E), σ_G^2 (thus σ_G), and r_{GE}, for each combination of values. He claims that when restricted by the reasonable assumption that all estimated variance components (as well as r_{GE}) must be positive, "then *only one* of the entire 60 sets [combinations] of equations yields a *realistic* solution" (1976, p. 98). That combination was .70 for $r_{E1E2(MZT)}$, .70 for $r_{E1E2(DZT)}$, and .50 for $r_{G1G2(DZT)}$. This combination produces the following solution: $\sigma_E^2 = 60.00; \sigma_G^2 = 139.50;$ and $r_{GE} = .078.$ Translated into proportions of explained variance, this gives:

$$h^2 = \frac{\sigma_G^2}{\sigma_Y^2} = 139.50/213.75 = .65 \text{ (heritability)}$$

$$e^2 = \frac{\sigma_E^2}{\sigma_Y^2} = 60.00/213.75 = .28 \text{ (environmentability)}$$

$$2her_{GE} = 2[\sqrt{.65}\sqrt{.28}(.078)] = .07 \text{ (proportion explained by gene-environment covariance)}$$

Jensen claims that since only this combination yields a permissible and "realistic" solution (one wherein all variance components, including r_{GE}, are positive), this constitutes evidence that heritability is greater than environmentability and that MZTs and DZTs probably do not differ in their respective degree of environmental similarity, since the one admissible solution had assigned equal values (.70) to both $r_{E1E2(MZT)}$ and $r_{E1E2(DZT)}$. But as Goldberger and Lewontin (1976) have noted, there are additional combinations of Jensen's own hypothesized parameter values that yield admissible (and certainly realistic) solutions. Jensen's calculations are simply incorrect. For example, as one can easily verify, setting $r_{E1E2(MZT)}$ at .90, $r_{E1E2(DZT)}$ at .60, and $r_{G1G2(DZ)}$ at .50, the following admissible solution is obtained: $\sigma_E^2 = 180.00; \sigma_G^2 = 31.50;$ and $r_{GE} = .0149.$[1] Translated into proportions, this gives

$$h^2 = 31.50/213.75 = .15$$
$$e^2 = 180.00/213.75 = .84$$
$$2her_{GE} = 2[\sqrt{.15}\sqrt{.84}(.0149)] = .01,$$

a solution in which e^2 is considerably greater than h^2, thus completely reversing Jensen's conclusion. The assigned values for the environmental correlation for MZTs (.90) are greater than the assigned environmental correlation for DZTs (.60), certainly more plausible than equal environmental correlations.

Still another combination of Jensen's hypothesized values yields an admissible solution: Setting $r_{E1E2(MZT)} = .80$, $r_{E1E2(DZT)} = .60$, and $r_{G1G2(DZ)} = .58$, the following solution is obtained: $h^2 = .58$, $e^2 = .42$, and $2her_{GE} = 0$. Jensen's claim of finding "only one realistic" possible solution is simply arithmetically wrong.

It appears that Jensen's algebra is also lacking. As Goldberger and Lewontin (1976, pp. 4–8; cf. Goldberger, 1976) have shown, there is evidently a fairly wide range of values for the two environmental correlations and the genetic correlation that yields admissible solutions. Even sticking with Jensen's restrictions that $.70 \leq r_{E1E2(MZT)} \leq .90$ and that $.60 \leq r_{E1E2(DZT)} \leq .90$, then allowing $r_{G1G2(DZ)}$ to vary from .28 to .59 (somewhat less restrictive than Jensen's hypothesized values, but still conceivable even for genetic theory), h^2 can go from .16 to .72; e^2 from .28 to .84; and r_{GE} from zero through .25. These results are obtained using Jensen's observed IQ correlations (.87 for MZTs and .56 for DZTs) borrowed from Erlenmeyer-Kimling and Jarvik. Even slightly different values for the IQ correlations will produce still wider fluctuations in the resulting values of h^2, e^2, and r_{GE}. (Part of Jensen's analysis in Jensen, 1975, 1976, incidentally, is based entirely on an earlier one by Hogarth, 1974, even to the point of using precisely Hogarth's values for Δr_{E1E2} [from zero to .20] and for $r_{G1G2(DZ)}$ [.50, .54 and .58]. Jensen, 1976, p. 97, gives the impression that he performed an analysis leading to three figures he presents on p. 97, but in fact the analysis and figures are from Hogarth, 1974.)

The heart of the problem is of course that there are too many unknowns in any system of kinship equations to permit meaningful solution. Only slight variations in parameter values can produce values of both h^2 and e^2 that go virtually from zero to 1.00, which is hardly useful information about the relative influence of genes and environment upon IQ. Essentially Jensen is trying to estimate one unknown (h^2) from a single kind of datum (the observed correlations, r_{Y1Y2}) while at the same time $r_{E1E2(MZT)}$, $r_{E1E2(DZT)}$, $r_{G1G2(DZ)}$, σ_E^2 (and thus e^2), and r_{GE} are all unknown. Furthermore, as already noted, Jensen's equations (5.1)–(5.3) and (5.6)–(5.8) arbitrarily assume that r_{GE} is the same in the general population, for MZTs, and for DZTs. These equations also assume, though Jensen does not so state, that the cross-pair gene-environment correlations are equal, such that $r_{G1E2} = r_{G2E1}$ for each of the kinship categories used. The equations also assume that all variance resulting from measurement error is removed by the attenuation correction, thus ignoring any possibility of systematic measurement bias. They assume that all other assumptions used to get equation *(4.14)* are correct—additivity, linearity, recursiveness, uncorrelated residual (nonfamily) environments, no direct effects of the IQ score of one pair member upon the IQ score of the other, and so on.

THE HONOLULU RESEARCH

Having shown that pairwise comparisons of kinships and kinship comparisons involving the simultaneous solution of three equations are certainly not promising, the question now arises whether working with a larger number of kinships (including direct line or parent-offspring kinships) might not yield more insight. Since each new equation introduces several new unknowns into the system, additional assumptions must be introduced to solve the system. I remind the reader again that assumptions must be made in any scientific empirical analysis. I make them myself when I critically evaluate the models of others. My quarrel is that a significant number of the important assumptions are arbitrary or implausible or both.

The Population Genetics Laboratory at the University of Hawaii at Honolulu has used this approach. I draw mainly upon three important works here: Rao et al. (1976, 1978) and Rao and Morton (1978). I also rely in part on the extensive and exceedingly useful critical discussions of these works by Goldberger (especially 1978b and also 1978c, 1978d), as well as on additional Honolulu publications (for example, Morton, 1974; Morton and Chung, 1978; Morton and Rao, 1978; Rao et al., 1974). The essential art of the sophisticated Honolulu research (true also of the Birmingham research mentioned below) is to formulate equations, one for each kin, so that the number of equations is greater than the total number of unknowns to be estimated. This ideal condition of overdetermination (overidentification) permits—with many assumptions and constraints—solution to the various unknowns.

A distinction the Honolulu group has reintroduced into the literature is that between adult and child heritability, and between adult and child environmentability. The distinction was first made nearly fifty years ago by Sewell Wright himself (1931, cited in Morton, 1972) in a path analysis of data from Burks (1928)—which, incidentally, itself contained one of the earlier applications of path analysis to anything

(Burks did a path analysis of parental IQ and parental environment on child's IQ, and investigated covariance between parental IQ and environment). Separate genetic and environmental path coefficients were used for different generations. The Honolulu researchers, like Wright (1931), have consistently found IQ heritability for adults to be less than IQ heritability for children, which they attribute to "the leveling effect of the school system [which] is replaced by varying stimulation in different occupations" (Rao et al., 1976, p. 238). (Despite this, Morton, 1974, p. 327 had remarked earlier that one "would be quite unjustified in claiming that heritability is relevant to educational strategy.") They find adult heritability (q^2, the notation employed here, after Goldberger, 1978b) to be around .30 and child heritability (h^2) to be around .70. They find adult environmentability (p^2) to be about .50 and child environmentability (they use the symbol c^2 rather than e^2; c^2 is used in this chapter) to be quite low, around .10 or .15. Remaining IQ variance is then allocated to gene-environment correlation (in Rao et al., 1976, but not in Rao and Morton, 1978) and to residual (noncommon or specific) environment.

In Rao et al. (1976, pp. 237–38) they find that $h^2 = .670$ and $c^2 = .094$; and that $q^2 = .211$ and $p^2 = .506$. Goldberger (1978c, 1978d), however, discovered a specification error in their equation *(8)* in Figure 5.1, which when corrected [2] by them (in Rao et al., 1978) resulted in the following slightly different estimates: $h^2 = .711$; $c^2 = .084$; $q^2 = .320$; and $p^2 = .499$. (Their correction also entailed treating the Newman et al., 1937, separated identical twins as adults, which for the most part they were, instead of as children, as they had done originally. This meant using q^2 rather than h^2 in equation [2] in Figure 5.1, which, as it turns out, removed an indeterminacy in their model that was created by the error discovered by Goldberger.) Gene-environment correlation, their correlation *a*, assumed to be the same for adults and children across generations as well as across kinships, was estimated by them to be .174, thus allocating 2*cha* or .085, or 8.5 percent, as the amount of the total IQ variance attributable to childhood gene-environment covariance. This leaves $1 - (h^2 + c^2 + 2cha)$ or .12 (12 percent of the IQ variance) to residual childhood environment. For adults, the portion of the total variance attributable to gene-environment covariance was 2*pqa* or .139. Thus, $1 - (q^2 + p^2 + 2pqa)$ comes out to .042, or 4.2 percent, of the IQ variance attributable to residual adult environment.

In Rao and Morton (1978, p. 159), $h^2 = .689$; $c^2 = .157$; $q^2 = .301$; and $p^2 = .549$, roughly comparable to the 1976 estimates. But in the later analysis, they assume a to be zero (as a consequence of another assumption, namely that their path coefficient $x = 0$ in Figure 5.2; see note a in Table 5.1); hence $2cha = 0$ and $1 - (h^2 + c^2) = .154$ or 15.4 percent for childhood residual environment. Similarly, for adults, $1 - (q^2 + p^2) = .15$ or 15 percent.

The Honolulu analyses are based on the path diagrams given here in Figure 5.1 (for collateral pairs) and Figure 5.2 (for nuclear families, that is, direct line kinships, and adoptions). Figure 5.2, the heart of the Honolulu research, carries by itself all the necessary information and is central to all their analyses. Figure 5.1 is heuristically useful but not necessary for their strategy. Figure 5.1 is virtually identical to Figure 4.2; it is a generalized additive path model for the sources of IQ similarity for collateral pairs of individuals. In their model, the measured IQ scores (not the "intelligence") of each pair member are X and Y (corresponding to our Y_1 and Y_2); G is genotype; and C is common environment, intended to represent childhood environment even on data for adults. As subscripts, X and Y correspond to pair members 1 and 2. The I (for index) variables are regarded as indexes of environment, such as socioeconomic status. A flaw in their model and analysis is their assumption that socioeconomic status validly measures environmental construct C in Figures 5.1 and 5.2. Even assuming they can measure socioeconomic status perfectly (they use the Burks 1928 culture index for their 1976 analysis and correlations based on indexes used by Duncan et al., 1972, Duncan and Featherman, 1973, and Scarr and Weinberg, 1977, for their 1978 analysis), no sociologist would ever argue that SES captures all or even most of even the common social environment, C. Furthermore, the path coefficient i, in effect a *measurement coefficient* for their environmental index, consistently comes out to be implausibly high, at times reaching .999 and even 1.00.

The phenotypic variables (X, Y) and environmental indexes (I_x, I_y) are of course treated as measured (observed) variables, while constructs G_x, G_y, C_x, and C_y are unmeasured (unobserved). Path coefficients h_x, h_y, c_x, and c_y correspond to our h_1, h_2, e_1, and e_2. The gene-environment correlations a_x, a_y, t^*, and s^* correspond to our r_{G1E1}, r_{G2E2}, r_{G1E2}, and r_{G2E1}. Correlation m^* is our genetic correlation r_{G1G2}; and correlation u^* is the environmental similarity correlation, r_{E1E2}. Residual variables are present for phenotypic dependent variables X and Y, but the Honolulu researchers prefer not to state them explicitly in the path diagram.

Figure 5.2 represents the Honolulu conceptualization of causation for direct line, that is intergenerational, kinships. It is also used for

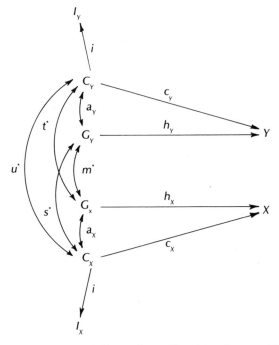

Figure 5.1. The Honolulu model for collateral kinships. *Source:* After Rao et al., 1976, p. 231. *Note:* G is genotype; C is common environment; I is environmental index; X and Y are phenotypes (IQ scores) of the two pair members. Subscripts X and Y refer to the pair members. In the original diagram, measured variables are in boxes (these are phenotypes X and Y, and indexes I_x and I_y), and unmeasured constructs are in circles (G_x, G_y, C_x, and C_y). Each phenotype (X, Y) has its own residual, although Rao et al. (1976) do not explicitly state them in their path diagram.

adoptions. It is almost identical to the path diagram used by Jencks et al. in their analysis of parent-offspring correlations and adoptions (1972, p. 268). (The Honolulu model provides for a path [f] from parental environments C_M and C_F to childhood environment C_C, while the Jencks et al. model does not. Otherwise, except for separately designating adult and child heritability and environmentability coefficients, the models are the same.) The subscripts M, F, and C refer to mother, father, and child; P_M, P_F, and P_C are the measured IQ scores (Ps) of each. Constructs G_M and G_F are the parental genotypes; C_M and C_F are the *childhood* environments of the now adult parents, presumed to be measured by indexes I_M and I_F respectively, with i as their

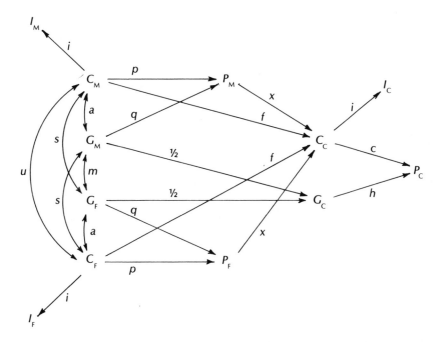

Figure 5.2. The Honolulu model for nuclear families (direct line kinships and adoptions). *Source:* After Rao et al., 1976, p. 229 and Rao and Morton, 1978, p. 146. *Note:* The path coefficients q and p (after Goldberger, 1978b) appear in place of the original investigators' hz and cy respectively. G is genotype, P is phenotype (IQ score), and C is common childhood environment, with index I. Subscripts F, M, and C represent father, mother, and child. In the original diagram, measured variables (all P and I) are in boxes; unmeasured constructs (all G and C) are in circles. Each phenotypic variable (P_F, P_M, P_C) has its own residual, as do C_C and G_C, though they are not stated in the diagram. In Rao and Morton (1978, p. 146), the path coefficient f is distinguished for father and mother as f_f and f_M respectively, as are coefficients x (as x_F and x_M) and i (as i_F and i_M), although equality of f_f to f_M, x_F to x_M, and i_F to i_M is assumed in actual analysis.

path coefficient. (In Rao and Morton, 1978, parental and child coefficients for environmental indexes are distinguished as i_M [$= i_f$] for parents and i_C for each child.) Path coefficients h and c, and correlations a, are as defined in Figure 5.1. New path coefficients are q, the effect of adult genotype on adult IQ score; p, the effect of the adult's childhood environment on his or her adult IQ score; a coefficient (always preset at $\frac{1}{2}$) for the effect (transmission) of parental genotypes G_M and G_F to child genotype G_C (a coefficient that Jencks et al., 1972, p. 281, allow to vary between .25 and .50, which means that the Jencks et al. heritability estimate comes out to .29 if this genetic transmission coefficient is set at .50); x, the effect of parental (as adult) IQ score upon own child's environment; and finally, path coefficient f, the effect of the parental (childhood) environment on own child's environment. Note that all path coefficients are shorthand notations such that $h = \beta_{P_C G_C}$; $c = \beta_{P_C C_C}$; $f = \beta_{C_C C_M} = \beta_{C_C C_F}$; and so on. They are by definition partial and not zero-order coefficients.

Correlations a represent the correlation between parent's genotype and that same person's childhood environment, separately for each parent and any given child. Note that what might be labeled as $a_M = a_F = a_C$, an a priori assumption in the model. The remaining correlations introduce new (though not unfamiliar) ideas: m is the presumed correlation between the genotypes of mother and father, that is, any presumed genotypic correlation arising from mating assortatively by IQ. (In Rao et al., 1976, 1978, and in Rao and Morton, 1978, m is assumed to be zero. For Jencks et al., 1972, p. 273, this coefficient comes out to be about .17.) The cross-pair gene-environment correlations (s) are assumed equal, so that $s_{C_M G_F} = s_{C_F G_M} = s$. In their analyses, the Honolulu researchers assume further that these correlations are zero, so that $s_{C_M G_F} = s_{C_F G_M} = s = 0$. Correlation u represents the correlation between the childhood family environments (as constructs, not indexes) of adult spouses. Finally, unstated residuals appear not only for the three phenotypic variables P_M, P_F, and P_C, but for child environment C_C as well. Genotype G_C also has a residual which can be taken to represent the Mendelian "segregation" from midparent (Goldberger, 1978, p. 41; Morton, 1974).

The Honolulu models thus assume additivity, linearity, and recursiveness. The gene-environment correlation a is assumed equal across generations, as are the cross-pair gene-environment correlations (s) for parents. But even if all parental gene-environment correlations a, s, and t are zero, if some assortative mating were to be assumed (thus $m > 0$), then via paths q, x, and genetic paths ($\frac{1}{2}$), it is *still* possible for childhood gene-environment correlation (a_C) to be nonzero. Furthermore, even if $m = 0$, then a_C could be nonzero via only paths q, x, and ($\frac{1}{2}$).

For collateral pairs (Figure 5.1), a_x is not necessarily assumed equal to a_y, nor is s^* to t^*. This feature is preserved in Rao et al.'s (1976) analysis, where (as in Jencks et al., 1972) gene-environment correlation is allowed to exist for adopted-natural pairs but not for adopted-adopted pairs (cf. equations [4] and [5] in Table 5.1). It is not preserved in Rao et al. (1978), however, since there, $a = 0$ for all collateral and direct line kinships (see Table 5.2).

From direct inspection of the path diagrams the following additional assumptions are apparent:

1. No residual variable is permitted to be correlated with any other variable (other than its own) including all other residual variables, in either model. Since residual variables are unstated, these assumptions can be regarded as implicit. A large number of correlations must therefore be treated as zero. For example, nonfamily environment is not permitted a direct causal effect on spouses's nonfamily or family environment, on the nonfamily or family environment of one's child, or on the IQ score of one's child. The reader is invited to locate other similar and similarly implausible inferences.

2. No causal paths emerge from any of the environmental indexes. Nor, given recursiveness, do any paths end at any index in either model. Some highly suspect hypotheses thus necessarily arise. For example, a person's childhood SES (independently of other family environment effects) is not permitted to affect directly his or her adult IQ in either model.

3. The IQ score of one pair member (Figure 5.1) is permitted no direct causal effect upon the IQ score of the other pair member.

4. For collateral kin, $c_x = c_y = c$. Thus, asymmetric effects of prenatal and postnatal environment are not permitted.

5. The Honolulu researchers assume an imperfect measure of environment, by means of path coefficient i for the environmental indexes. They thus in principle allow for the possibility that this measure is less than perfect (that $i < 1$). But why assume that index I imperfectly measures common environment C, yet persist in having no construct for IQ score? What about that construct ("intelligence") of which IQ score (the "index") may be, however imperfectly, a measure? Should one wish to incorporate some such measurement coefficient, then all equations in Tables 5.1 and 5.2 would have to be rewritten. Furthermore, IQ is probably not a simple unidimensional index, thus implying the use of two or more "IQ" scores per individual. This then allows some nonrandom measurement error to enter (W from Chapter 4) which, if incorporated into the model, would again necessitate

Table 5.1. **Equations from Rao et al.: The Eleven-Kinship Set, Where** $m = s = 0$

Kinship category	n	Equation	Observed r	Rao et al. equation[a]	Modified equation
IQs of identical twins together (MZTXY)	50	1	.89	$.89 = h^2 + c^2 + 2cha$	$.89 = h^2 + u_1c^2 + 2cha$
IQs of separated identical twins (MZAXY)	19	2	.69	$.69 = \theta^2q^2$	$.69 = \theta^2q^2$
IQs of siblings together (SSTXY, our SIBT)	2,001	3	.52	$.52 = \tfrac{1}{2}h^2 + c^2 + 2cha$	$.52 = \tfrac{1}{2}h^2 + u_3c^2 + 2cha$
IQs of adopted-adopted pairs (FSTXY, our UNT-AA)	21	4	.23	$.23 = \theta^2c^2$	$.23 = \theta^2u_4c^2$
IQs of adopted-natural pairs (FSPXY, our UNT-AN)	94	5	.26	$.26 = \theta(c^2 + cha)$	$.26 = \theta(u_5c^2 + cha)$
IQ of adopted child with his own index (OFPYIY)	186	6	.25	$.25 = \theta ic$	$.25 = \theta ic$
IQ of natural child with his own index (SSTXIX)	101	7	.44	$.44 = i(c + ha)$	$.44 = i(c + ha)$
IQ of parent with own child's index (OPTXIIY)	205	8	.69	$.69 = it$	$.69 = it$
IQs of adoptive parent and child (OFPXY)	1,181	9	.23	$.23 = \theta ct$	$.23 = \theta ct$

Table 5.1 (continued)

IQs of parent and child (OPTXY)	1,250	10	.48	$.48 = \frac{1}{2}hv + ct$	$.48 = \frac{1}{2}hv + ct$
IQs of spouses (FMTXY)	887	11	.50	$.50 = w$	$.50 = w$

Source: After Goldberger, 1978b, p. 74 and Rao et al., 1976, 1978.

[a] The 11 equations have 8 free parameters: h, c, q, i, f, p, x, and u_6, thus yielding 11(kinships) − 8(free parameters) or 3 degrees of freedom for chi square. The correlations u_1, u_3, u_4, and u_5 are preset. Goldberger (1978b, 1978c) notes that certain expressions can be written as derived parameters, with the following substitutions:

$$t = f[p(1 + u_6) + qa] + x(1 + p^2 u_6)$$
$$v = pa + q$$
$$w = p^2 u_6$$

Here u_6 is substituted for the Honolulu u simply to distinguish it. In the "control" analysis I do below, u_6 is treated as a free parameter, as in the Honolulu formulation. Derived parameters a and θ can be defined from Rao et al. (1976, pp. 229–30; Goldberger, 1978b, p. 52) as follows:

$$\theta = (1 - 2cha)^{-\frac{1}{2}} = 1/(1 - 2cha)^{\frac{1}{2}}$$
$$a = qx/(1 - f - px).$$

The expression θ represents the ratio of the IQ standard deviation for children raised by their own natural parents to that for adopted children.

Table 5.2. Equations from Rao and Morton: The Sixteen-Kinship Set, Where $m = s = x = 0$, Hence $a = 0$ and $\theta = 1$

Kinship category	n	Equation	Observed r	Rao and Morton equation[a]	Modified equation
IQs of identical twins together (MZTXY)	421	1	.842	$.842 = h^2 + c^2$	$.842 = h^2 + u_1 c^2$
IQs of separated identical twins (MZAXY)	19	2	.679	$.679 = q^2$	$.679 = q^2$
IQs of siblings together (SSTXY, our SIBT)	2,467	3	.516	$.516 = \tfrac{1}{2}h^2 + c^2$	$.516 = \tfrac{1}{2}h^2 + u_3 c^2$
IQs of adopted-adopted pairs (FSTXY, our UNT-AA)	421	4	.360	$.360 = c^2$	$.360 = u_4 c^2$
IQs of adopted-natural pairs (FSPXY, our UNT-AN)	228	5	.283	$.283 = c^2$	$.283 = u_5 c^2$
IQ of adopted child with his own index (OFPYIY)	774	6	.286	$.286 = ic$	$.286 = ic$
IQ of natural child with his own index (SSTXIX)	4,717	7	.304	$.304 = ic$	$.304 = ic$
IQ of parent with own child's index (OPTXIY)	1,272	8	.570	$.570 = it$	$.570 = it$
IQs of adoptive parent and child (OFPXY)	1,181	9	.228	$.228 = ct$	$.228 = ct$

Table 5.2 (continued)

IQs of parent and child (OPTXY)	1,310	10	.484	$.484 = \frac{1}{2}hq + ct$	$.484 = \frac{1}{2}hq + ct$
IQs of spouses (FMTXY)	1,118	11	.511	$.511 = p^2u$	$.511 = p^2u_6$
Indexes of spouses (FMTIXIY)	1,165	12	.226	$.226 = i^2u$	$.226 = i^2u_6$
Indexes of natural parent and child (OPTIXIY)	17,432	13	.343	$.343 = ijt/p$	$.343 = ijt/p$
IQ of parent with his own index (OPTXIX)	887	14	.347	$.347 = ip$	$.347 = ip$
IQs of natural parent and adopted-away child (OPAXY)	63	15	.407	$.407 = \frac{1}{2}hq$	$.407 = \frac{1}{2}hq$
IQs of separated siblings (SSAXY)	125	16	.249	$.249 = \frac{1}{2}h^2$	$.249 = \frac{1}{2}h^2$

Source: After Goldberger, 1978b, pp. 76–78, and Rao and Morton, 1978, p. 156.

[a] The 16 equations have 8 free parameters: h, c, q, i, t, p, j, and u_6. The correlations u_1, u_3, u_4, and u_5, are preset. The only free parameter not in Table 5.1 is path coefficient j, representing a path from common parental environment to its index. This is treated as distinct from the path (i) from child environment to its own index. There are thus 16(kinships) − 8(free parameters) = 8 degrees of freedom for chi square. The derived parameter t is $t = fp(1 + u_6)$, since now $a = x = 0$. Note also that v and w are eliminated because $v = q$ since $a = 0$; and $w = p^2u_6$ which is equation (11) as stated. Note finally that since $a = 0$, the gene-environment covariance quantity $2cha = 0$, and thus no longer appears in any equation.

rewriting all the equations for both analyses (Tables 5.1 and 5.2).
6. The index used for the 1976 analysis (Rao et al., 1976, 1978) is a primitive index from Burks (1928), consisting of a set of 5-point scales referring to parent's speech, education, interests, home library, and artistic tastes. Goldberger (1976b, 1978b) notes that for Burks's "culture index," removing family portraits from the walls and jazz from the record collection would raise a family's score by as much as attending college for four years! Such an index is hardly representative of the SES indexes used in modern-day sociology. The Honolulu researchers, who have berated social scientists for straying into the field of behavioral genetics (as in Rao and Morton, 1978, p. 172), might profitably note the advice of sociologists when they stray into sociology. Fortunately, they do employ correlations for environmental indexes from Duncan et al. (1972) and Duncan and Featherman (1973) in their 1978 data set and analysis (Rao and Morton, 1978).

Despite these shortcomings, the Honolulu approach has certain important advantages over past analyses. Among these are:
1. Specification of different path coefficients for adult and child heritability and environmentability.
2. Specification of a causal path (f) from parental environments C_M and C_F to the children's environment, C_C.
3. Partitioning environmental variance (c^2 for children; p^2 for adults) from residual environmental variance. This permits the calculation of both sources of environmental variance combined, such that the total environmental effect is: $[c^2 + $ (child residual)$^2]$ for children, and $[p^2 + $ (adult residual)$^2]$ for adults. This means that in general,

$$1 = h^2 + c^2 + 2cha + \text{(child residual)}^2$$

and

$$1 = q^2 + p^2 + 2pqa + \text{(adult residual)}^2.$$

Recall that Jencks et al. sometimes implausibly assumed that (heritability) + (environmentability) + (the covariance quantity) = 1 (1972, p. 269; cf. Taylor, 1973b).
4. The single most important advantage of the Honolulu approach (shared with the Birmingham group discussed later) is that they attempt to fit a full set of equations to a full set of data, that is, to different observed IQ correlations, one for each kin. Goodness of fit for the resulting expected correlations ($\hat{r}$, calculated from the

estimated parameters h, c, p, q, etc.) is then assessed by a chi-square test with N(kinships) $-$ k(unknown free parameters) degrees of freedom. Since $N > k$, the equation system is desirably overdetermined. Weighted least squares procedures are then used to estimate the unknown parameters, where the number of pairs for each observed r are explicitly taken into account. This procedure is considerably more systematic and consistent with contemporary statistical theory and practice (see, for example, Jöreskog, 1973) than is, say, the Jencks et al. (1972) imaginative but tortuous attempt to get various parameter estimates more or less to agree with each other via various pairwise comparisons among many different kinds and forms of equations. (Jencks et al.'s system of comparing equations borders on the unfathomable, as Goldberger notes, 1978a, p. 10. When Jencks et al. were preparing their study, however, the use of weighted least squares estimation of overdetermined systems was quite new to the social and behavioral sciences.) The Honolulu approach is clearly superior to Jensen's (1976) abortive attempt to solve three equations for unique parameter values.

The Honolulu Equations and a Modification

The Honolulu equations are displayed in Table 5.1 (the 1976 11-kin analysis and data set) and Table 5.2 (the 1978 16-kin analysis and data set). The first column in each table describes the particular kinship category. The second column gives the number of pairs used to calculate the observed kinship correlation r (given in the fourth column), a weighted average correlation based on different studies of the particular kinship. The correlations for the 1976 analysis are drawn from the Jencks et al. (1972) compilation with various correlations from Burks (1928) also included. The correlations for the 1978 analysis are taken from these two sources and from Duncan et al. (1972), Duncan and Featherman (1973), and Scarr and Weinberg (1977). The observed correlations are listed in Tables 5.1 and 5.2 as given in Goldberger (1978b, pp. 50, 76), who presents the correlations untransformed by the z-transformation.

The actual studies used in calculating the averaged correlations shown are not listed here, but there are important flaws in the Honolulu group's selection and use of studies noted by Kamin (personal communication, 1979) and by Goldberger (1978b). Some of these problems are discussed later. More pairs per kin are used in the 1978 data set (Table 5.2) than in the 1976 data set (Table 5.1), so that the observed rs differ somewhat. Five kinships (equations [12]–[16]) were added for the 1978b analysis. The equations are given in the fifth column of each table and numbered in the third column, using Goldberger's (1978, pp. 74, 78) symbols (substituting the single coefficient p for their cy, q for their hz, and j for their i_M [$= i_f$] indexes).

Complete and lucid derivations of each Honolulu equation in both tables appear in Goldberger (1978b). Each equation in Table 5.1 can be obtained from direct application of the path theorem while applying the constraints given in Rao et al. (1976, pp. 232–33, their Table 4) and Rao et al. (1978) and using Goldberger's (1978b, pp. 52–59) definitions of derived parameters t, v, and w (given here in Table 5.1, note a). These constraints pertain to such matters as when $a = 0$ and when it does not (for example an individual's own $a = 0$ if that individual is adopted and if no selective placement is assumed, as with the Honolulu 1976 and 1978 approaches); when child or adult path coefficients h or q and c or p are to be used in an equation; the value of m^* (.5 for sibling pairs; but $m = 0$ for married pairs); and where θ is used. This quantity, θ, is the ratio of the IQ standard deviation for children reared by their own parents to that for adopted children; a particular correlation is multiplied by θ if one pair member is adopted, and by θ^2 if both are adopted, and thus θ is a correction for restriction of range of the total IQ variance of all adopted pairs. For the 1976 analysis (Table 5.1), it was assumed that $m = s = 0$. As a consequence of finding a low value for path x (.179 in Rao et al., 1978, p. 447), which is the effect of parental IQ on child environment, it is assumed that $m = s = x = 0$ for the 1978 analysis (Table 5.2). Thus given the definition of θ and a (Table 5.1, note a), $\theta = 1$ and $a = 0$; and no gene-environment correlation in any equation appears in the 1978 16-kin analysis (see Table 5.2).

Using the path theorem with these constraints, examples of how these equations can be derived are given here. Equation (6) in Table 5.1 (from Figure 5.2) is $.25 = ic$ direct from the path theorem ($a = 0$ since the child in question is adopted); multiplication of the correlation by θ gives $.25 = \theta ic$. Thus $.286 = ic$ in Table 5.2 (since now $\theta = 1$). For equation (7) in Table 5.1 (from Figure 5.2 where $a \geq 0$), the path theorem gives

$$.44 = ic + iah$$
$$= i(c + ha),$$

and hence $.304 = ic$ in Table 5.2 since $a = 0$. Note that equations (6) and (7) in Table 5.2 (not Table 5.1) are thus identical, and therefore will give the same predicted correlation ($\hat{r}$, not shown) after arriving at the solved values for i and c. Note finally that equations *(1)–(5)* in both Tables 5.1 and 5.2 are collateral kin and thus represent derivations from the path theorem examined in Chapter 4. For example, equation *(4.28)* for MZTs and equation *(4.30)* for SIBTs are the same as Honolulu equations *(1)* and *(3)* respectively in Table 5.1 if one assumes (as do the Honolulu models) that MZTs and SIBTs do not differ in environmental correlation u^*.

This realization makes it possible to take a detailed look at a crucial Honolulu assumption that represents an important weakness in their model: For each raised-together collateral kin (equations [1] and [3]–[5]), environmental similarity (u^*, represented from here on as u or as u_i for any given kin, i) is assumed equal (thus $u_i = 1$ in effect) across all four of these kinships, including identical twins raised together, biological siblings raised together, adopted-adopted pairs, and adopted-natural pairs raised together. The implausibility of this assumption in view of the available empirical evidence has been stressed throughout this book. The Honolulu specification that $u = 1$ by definition since the C_x and C_y represent the environment common to a pair is an obfuscation (Rao et al., 1976, p. 229 and passim; Rao and Morton, 1978, p. 148). It formulates the matter of intrapair environmental correlation in such a way that the issue does not in effect arise. Since their solutions come out to $u = .9996$ (in Rao et al., 1978, p. 447) and $u = .940$ (in Rao and Morton, 1978, p. 158), their model and specification yield little difference in environmental correlation between all collateral pairs and the childhood environments of adult married pairs as well. This means concluding that the childhood environments of now-married adults are as similar to one another as are, say, the environments of identical twins raised together from birth in the same home! Their specification even means concluding that the childhood environments of a now-married adult couple are as similar as the environment of a single individual correlated with itself.

The implausibility of assuming equal environmental correlations for collateral kinships *(1)* and *(3)–(5)*, and the strong suspicion about any model specification that yields such high values for married pairs, suggests that the Honolulu equations might profitably be respecified to incorporate varying u values for at least the collateral kin. This is done in the last column of Tables 5.1 and 5.2. For these modified Honolulu equations, some u_i (from the path theorem, Figure 5.1, and Figure 4.2) is inserted. This is done by stating (in Tables 5.1 and 5.2) different u_i values (designated u_1, u_3, u_4, and u_5) for the raised-together collateral kinships, and one value (u_6) for other kinship categories. (This causes

u_6 to be equal across equations [8]–[11] in Figure 5.1 and equations [8]–[13] in Figure 5.2, as with the original Honolulu equations.) Different sets of arbitrary numerical values are assigned to u_1, u_3, u_4, and u_5 (first leaving u_6 as a free parameter, as with the Honolulu analyses, and then constraining it to some value), and the resulting solutions for parameters of interest (particularly h^2, c^2, q^2, and p^2) are observed.

Such an analysis, while admittedly only suggestive (the assigned u_i values are arbitrary), certainly gives some idea of how very sensitive the genetic and environmental parameters are to different environmental correlations within the framework of the model. Allowing for different correlations in family environments of collateral kinships is not the same as attributing differences in total variance to possible differences in residual (specific) environments, as done by Jencks et al. (1972, pp. 297–301) and also by Loehlin in a recent reanalysis of the Jencks et al. correlations (1978, p. 429; cf. p. 420).

There are other suspect assumptions pertaining to the environmental correlation quantity underlying the Honolulu equations for both data sets. I mention four of the important ones here:

1. No parameter for environmental correlation appears in the equation for the Newman et al. 19 pairs of separated identical twins (equation [2] in Tables 5.1 and 5.2). Thus the u_i for separated MZ twins (our u_2) is assumed to be zero, and these separated identical twins are presumed to have no environmental similarities. (As demonstrated in Chapter 3, this is not the case.) Even the Jencks et al. analysis of the Newman et al. separated twins allowed for a nonzero environmental correlation. Furthermore, Kamin (1974) certainly alerted the Honolulu researchers to the implausibility of the assumption. They also assume zero gene-environment correlation in both the 1976 and the 1978 analyses in the equation for separated MZs. In place of their equation (2) in Table 5.1, a fairer specification might be

$$.69 = \theta^2(q^2 + u_2c^2 + 2cha),$$

where u_2 might be estimated from correlations between SES and educational characteristics of adopting and natural parents, as done by Jencks et al., although admittedly given the small n_i (19), such a specification might not have a large effect on the resulting parameter solutions—depending of course on how many equations from the full set one wishes to solve in any given analysis.

2. They explicitly assume throughout that there is no selective

placement of adopted children, and thus no correlation between some environmental variable of the natural parents and the same environmental variable for the adopting parents. Yet there is direct evidence in the studies they cite that the educations of natural and adopting parents do indeed correlate—at about .29, a weighted correlation after Jencks et al. (1972, p. 278) based on Leahy (1932, 1935) and Skodak and Skeels (1949); or about .22 based on Scarr and Weinberg (1976, 1977, p. 184). This would produce a nonzero r for any correlation between the C (environment) of either natural parent and the C for either adopting parent, and thus would have to enter several of the Honolulu equations, such as equations (2), (5), (6), and (9) in Tables 5.1 and 5.2 and equations (15) and (16) in Table 5.2. (Loehlin's, 1978, analysis does permit a nonzero correlation for selective placement.)

3. Equation (16) in Table 5.2 represents the (collateral) category of separated siblings, based on the Freeman et al. (1928) study of 125 pairs. But Freeman et al.'s descriptive information is insufficient to determine how "separated" these siblings actually were. Should any environmental correlation (call it u_{16}) have characterized these separated sibling pairs, then equation (16) would have to be respecified as:

$$.249 = \tfrac{1}{2}h^2 + u_{16}c^2,$$

even with an assumption that $a = 0$.

4. Because of the equal-environment assumption and the 1978 assumption that $a = 0$, certain of the Honolulu equations for the 1978 analysis reduce to a single Jensen-style equation where heritability is a simple function of the difference between the phenotypic correlation of two kinships. For example, subtracting equation (3) from equation (1), equation (4) from equation (1), equation (5) from equation (1), equation (4) from equation (3), and equation (5) from equation (3) (Table 5.2) yields,

$$h^2 = (r_{MZTXY} - r_{SSTXY})/.5 = 2(r_{MZTXY} - r_{SSTXY});$$
$$h^2 = (r_{MZTXY} - r_{FSTXY});$$
$$h^2 = (r_{MZTXY} - r_{FSPXY});$$
$$h^2 = (r_{SSTXY} - r_{FSTXY})/.5 = 2(r_{SSTXY} - r_{FSTXY}); \text{ and}$$
$$h^2 = (r_{SSTXY} - r_{FSPXY})/.5 = 2(r_{SSTXY} - r_{FSPXY}).$$

Each of these is simply an exact restatement of the Jensen (1967) equation, the flaws of which have already been discussed at great length. For all their sophistication, the Honolulu research-

Table 5.3. **Solutions to Honolulu Equations** *(1)–(5)* **from Table 5.1 (1976 Data Set, Where** $m = s = 0$**) with Alternative Values for the Sibling Correlation**

	Observed r for siblings (SSTXY) or fraternal twins (DZTXY)		
	.52	.63	.70
h^2	.734	.536	.446
c^2	.372	.127	.101
q^2	.839	.530	.438
a	−.208	.444	.860
$2cha$	−.217	.232	.365
χ_1^2	.35	.53	6.01
P	>.50	>.30	<.02

ers have not moved far from the Jensen (1967) procedures, at least in the study of collateral kin.

An Illustrative Analysis

I look briefly now at what can happen to the Honolulu estimates of genetic and environmental parameters when one introduces two plausible changes in the data and one plausible change (involving the u_i) in the model. The results show how sensitive the childhood parameters h^2 and c^2 are to certain modifications of equations *(1)*, *(3)*, *(4)*, and *(5)*. The procedure employed is iterative weighted least squares, using a program developed at Princeton University by Goldfeld and Quandt (1972) called GQ OPT, which uses the computational algorithm given in Powell (1971). The input data consist of the observed correlations.[3] The basic idea in the least squares solutions here, as in the ones used by the Honolulu researchers and Goldberger (1978b), is to attempt to converge on unique values for the unknown free parameters while weighting the solutions on the basis of the number of pairs (the n_i sizes) for each observed correlation r_i. The goodness of fit of the observed correlations (r_i) to the predicted correlations $(\hat{r}_i)$ is assessed by chi square with N(equations) − k(free parameters) degrees of freedom. One seeks to minimize the differences between the observed and the expected correlations. (Expected correlations can be calculated directly by inserting the solved parameter values into the equations that contain them.)

I first explore two instances of how plausible changes in the data

Table 5.4. **Solutions to Honolulu Equations** *(1)–(5)* **from Table 5.1 (1976 Data Set, Where** $m = s = 0$**) with Alternative Values for the Separated MZ Twin Correlation**

	Observed r for separated identical twins (MZAXY)			
	.69	.50	.40	.30
h^2	.734	.733	.731	.731
c^2	.289	.457	.533	.604
a	−.145	−.262	−.303	−.338
$2cha$	−.133	−.303	−.378	−.449
χ_2^2	.58	.59	1.20	2.09
p	>.70	>.70	>.70	>.30

can change the parameter estimates for only the collateral equations *(1)* through *(5)*. Note from the 1976 equations in Table 5.1 that no equation appears for fraternal twins raised together (DZTs) but only one for siblings (equation 3). The reason is apparent enough: For the Honolulu formulation, DZTs would be treated exactly the same as siblings, and elimination of DZTs eliminates a redundancy in their model. But evidence exists (reviewed in Chapter 2) that assuming $a = 0$, DZTs and siblings differ in environmental correlation (u_i). Since heritability decreases as the observed difference between MZTs and DZTs decreases, one can ask how h^2 and even q^2 (as well as c^2) might be affected if an observed correlation for DZTs is put in place of the .52 for siblings.

This is done in Table 5.3. Values of .52, .63, and .70 are inserted as the observed *r* for equation *(3)* (with the n_i unchanged and the u_i set $= 1$ as with the Honolulu equations). Equations *(1)* through *(5)* are then solved simultaneously three times for these three IQ correlation values. The solution takes gene-environment correlation (a) as a free parameter, thus giving four free parameters that were solved ($h, c, q,$ and a), thus 1 degree of freedom. In an earlier article, Rao et al. (1974, p. 353) use an *r* of .63 ($n = 50$) for DZTs based on the Jencks et al. (1972, p. 286) reporting of the Newman et al. (1937) DZTs. Table 5.3 shows that h^2, q^2, and c^2 are quite sensitive to even the relatively slight differences in the observed DZT (sibling) correlation: As the observed correlation goes from .52 to .63, h^2 drops from .734 to .536, and q^2 from .839 to .530. Even c^2 decreases (from .372 to .127). Correlation a goes from negative to a more plausible positive, and the fit remains good

($\chi_1^2 = .53, P > .30$). The fit worsens (to $\chi_1^2 = 6.01, P < .02$) when the observed correlation is set at .70; h^2, q^2, and c^2 all drop further, but a becomes implausibly high (.860). I conclude that in this 5-equation (collateral kin) solution, both child (h^2) and adult (q^2) heritabilities as well as child environmentability (c^2) decrease quite markedly for even relatively slight increases in observed r for siblings (or DZTs). There is thus reason to be at least slightly skeptical of Morton's conjecture that "it is our contention, based on much experience, that perturbations of our data set have only trivial effects on the estimates" (1978, p. 193).

Table 5.4 shows some sensitivity of c^2 (but not h^2) to decreases in the observed r for separated MZ twins. Here, $\theta^2 q^2$ is replaced with $\theta^2 h^2$ in equation (2), Table 5.1, thus giving only three free parameters (h, c, and a). Based on the investigation in Chapter 3 (and hence the use of h^2 rather than q^2), as r is .69 (the Honolulu value), .50, .40, and .30, then c^2 goes from .289 to .604. The fit remains good, although a becomes increasingly negative. In Chapter 3 it was pointed out that any h^2 value greater than the observed correlation $r_{Y1Y2(MZA)}$ is highly suspect, which is the case here. There is thus reason to be suspicious of the specification of equations (1) through (5).

In Tables 5.5A and 5.5B the results obtained from 5-equation solutions to both Honolulu data sets are examined when the equations are modified as follows (see Tables 5.1 and 5.2): For the first ("control") solution, the environmental parameters u_1, u_3, u_4, and u_5 are all set equal to 1, thus preserving the Honolulu specification of equal environmental correlations across kinships. For the second solution (the second column of Table 5.5A), the u_i values are arbitrarily set at 1, .90, .85, and .80; for the third, at 1, .80, .73, and .65; and for the fourth, at 1, .70, .60, and .50.

As the differences among the u_i quantities increase, childhood heritability h^2 declines noticeably; adult heritability q^2 declines slightly; and childhood environmentability c^2 increases. Gene-environment correlation a (solved as a free parameter) remains low (though negative) and approaches zero—certainly a plausible outcome given the Honolulu hypothesis of low or near-zero gene-environment correlation. (The Honolulu researchers have also obtained low negative values for a.) The fit remains extremely good. Residual variance $1 - (h^2 + c^2 + 2cha)$ remains the same in all four analyses (.111, .113, .113, and .112), at a level quite consistent with the Honolulu findings. If one is willing to argue that the u_i values in the last column are even distantly plausible, then it is of particular interest that the h^2 of .417 is below the usual Honolulu value for child heritability of around .70,

Table 5.5. **Alternative Solutions to Modified Honolulu Equations** *(1)–(5)* **from Table 5.1 and Equations** *(1)–(5)* **from Table 5.2**

A. Using the 1976 data set, $m = s = 0$

$u_1 =$	1.0	1.0	1.0	1.0
$u_3 =$	1.0	.90	.80	.70
$u_4 =$	1.0	.85	.73	.60
$u_5 =$	1.0	.80	.65	.50
h^2	.734	.646	.543	.417
c^2	.372	.433	.473	.527
q^2	.839	.823	.730	.729
a	−.208	−.181	−.127	−.060
$2cha$	−.217	−.192	−.129	−.056
χ_1^2	.35	.42	.36	.27
P	>.50	>.50	>.50	>.50

B. Using the 1978 data set, $m = s = x = 0$ (hence $a = 0$)

$u_1 =$	1.0	1.0	1.0	1.0
$u_3 =$	1.0	.90	.80	.70
$u_4 =$	1.0	.85	.73	.60
$u_5 =$	1.0	.80	.65	.50
h^2	.567	.514	.430	.287
c^2	.249	.314	.402	.549
q^2	.679	.679	.679	.679
a	0.0	0.0	0.0	0.0
$2cha$	0.0	0.0	0.0	0.0
χ_2^2	9.44	8.30	4.93	1.34
P	<.01	<.02	>.10	>.50

and the c^2 of .527 is above the usual Honolulu value of around .10 or .15.

Table 5.5B, which shows the results of the same analysis on the Honolulu 1978 data and modified equations (from Table 5.2) and with correlation *a* set by them at zero, yields similar results: h^2 declines to .287, a long way from the Honolulu findings on childhood heritability; c^2 increases to .549, also quite far from Honolulu findings; and q^2 stays precisely the same. In general, the finding in Tables 5.5A and 5.5B of declining child heritability and increasing child environmentability is directly opposite to the persistent Honolulu result of high child heritability and low child environmentability. Once again considerable

Table 5.6. **Alternative Solutions to the Full Set of Modified Honolulu Equations** *(1)–(11)* **from Table 5.1 and Equations** *(1)–(16)* **from Table 5.2**

A. Using the 1976 data set, $m = s = 0$

$u_1 =$	1.0	1.0	1.0	1.0
$u_3 =$	1.0	.90	.80	.70
$u_4 =$	1.0	.85	.73	.60
$u_5 =$	1.0	.80	.65	.50
$u_6 =$	free	.75	.58	.40
h^2	.696	.696	.699	.706
c^2	.080	.083	.086	.090
q^2	.621	.629	.638	.651
p^2	.558(−)	.666(−)	.861(−)	1.248(−)
f^2	.023(−)	.020(−)	.018(−)	.017(−)
x^2	.147	.153	.156	.151
i^2	.815	.812	.812	.812
u_6	.805	(.750)	(.580)	(.400)
a	.211	.211	.211	.200
$2cha$	.100	.101	.104	.101
$2pqa$	−.248	−.273	−.313	−.361
χ^2	2.43	2.59	2.92	3.78
df	3	4	4	4
P	>.30	>.50	>.50	>.50

sensitivity of h^2 and c^2 to differences in u_i is seen. The residual $1 - (h^2 + c^2 + 0)$ again remains virtually unchanged (at .184, .172, .168, and .164).

Table 5.5B reveals an additional and particularly intriguing result: The Honolulu-style specification gives the worst fit of all four solutions ($\chi^2_2 = 9.44$, $P < .01$). The fit improves consistently (to $\chi^2_2 = 1.34$, $P > .50$) as the discrepancies among the u_i widen. I am reminded at this juncture of the following advice given by Rao and Morton after dismissing potential criticism of their specification of equal environmental similarity: "To such criticism there is only one answer: a statistical test of goodness-of-fit, which in samples of adequate size and structure can rule out an inadmissible model" (1978, p. 150).

The analyses thus far have involved only the collateral kin equations *(1)–(5)*. What happens if different values for the u_i (including u_6; see Tables 5.1 and 5.2) are inserted and solutions to the full set of equations are attempted? Tables 5.6A (for the 11-kin 1976 data set) and 5.6B (for the 16-kin 1978 data set) show the results. For each data set, four solutions were performed. In the first (control) solution for each data

Table 5.6 (continued)

B. Using the 1978 data set, $m = s = x = 0$ (hence $a = 0$)

$u_1 =$	1.0	1.0	1.0	1.0
$u_3 =$	1.0	.90	.80	.70
$u_4 =$	1.0	.85	.73	.60
$u_5 =$	1.0	.80	.65	.50
$u_6 =$	free	.75	.58	.40
h^2	.704	.714	.724	.734
c^2	.124	.113	.101	.091
q^2	.731	.771	.806	.836
p^2	.540	.654	.771	.884
f^2	.169	.170	.174	.190
j^2	.223	.250	.267	.275
i^2	.824	.896	.982	1.079
u_6	.901	(.750)	(.580)	(.400)
a	0.0	0.0	0.0	0.0
$2cha$	0.0	0.0	0.0	0.0
$2pqa$	0.0	0.0	0.0	0.0
χ^2	62.59	77.95	109.53	181.69
df	8	9	9	9
P	<.001	<.001	<.001	<.001

Note: Numbers in parentheses are the preset values for u_6.

set, u_6 is left as a free parameter and the other u_i are set at 1, thus approximating the Honolulu specification. The remaining three solutions set the u_i as shown. The results for the first solution for the 1976 data set, which include negative values for p and f, closely approximate those of Goldberger (1978, p. 75), but less so the results of Rao et al. (1978, pp. 446–47) which are the "corrected" results using $\theta^2 q^2$ in equation (2). The discrepancies are for p and f, which they find are .707 and .290 respectively. They evidently constrained p and f to be positive—though this constraint is not clearly stated (Rao et al., 1978). Goldberger notes that these negative values could be taken as evidence against the Honolulu specifications for the 1976 data set. The control solution for the 1978 data (Table 5.6B) shows positive values for all coefficients, and the estimates agree reasonably well with those of Rao and Morton (1978, p. 159). The fit is not at all good ($\chi^2_8 = 62.59$, $P < .001$), but their fit for the 1978 data for this particular specification was not that good, either. (Their χ^2_8 value was 39.83, $P < .001$. What they then did, as noted by Goldberger, 1978b, p. 77, was to redefine their test statistic as an F-statistic based on χ^2, df, and "error variance," and on this basis, they ended up not rejecting their model.) Overall, the coefficients remained quite unchanged as the u_i were varied, and the fit worsens considerably. One exception is p^2 (adult environmentability) in the 1978 analysis: It increases from .540 (for equal u_i values) to .884 (for the most discrepant u_i values). In general, changes in the environmental correlations have little effect on the parameters when the full set of equations is solved. This is not surprising, given that the u_i here pertain to the collateral kinships—for which there was considerable change in the solved parameters.

Some Conclusions

The examination of the Honolulu models and procedures permits the following conclusions:

1. Any single kinship equation will introduce several new unknowns into a system of equations, as Morton has noted before (1974, pp. 320–21). But to solve the resultant underidentified system, the Honolulu researchers have proceeded (as they must) by setting certain path coefficients and correlations at zero a priori, some explicitly and some not; and by assuming that certain effects (such as the environmental correlations and the gene-environment correlations) are equal across kinships. As a result, they end up with an overidentified system.
2. To achieve this overidentification, the Honolulu researchers make the following assumptions of questionable plausibility: that one's IQ score cannot affect how one is treated in one's own environment (via the recursiveness assumption); that pair members' IQ scores do not affect one another; that one's nonfamily environment cannot affect one's spouse's family environment, one's spouse's nonfamily environment, one's child's family environment, one's child's nonfamily environment, one's child's IQ score (all via the residual noncorrelation assumptions); that one's own SES cannot directly causally affect one's own IQ score; and that SES is a near-perfect indicator of "environment." A particularly disturbing assumption is that collateral pairs (genetically related or not) are not permitted to differ in their intrapair environmental similarities (the u_i). They also end up with the implausible empirical result that the correlation of the childhood environments of adult spouses nearly equals the correlation of the en-

vironments of identical twins raised together in the same home (as well as siblings raised together, adopted-natural pairs raised together, and adopted-adopted pairs raised together). Additional implausible restrictions include: no allowance for any degree of environmental correlation for separated identical twins, despite direct empirical evidence to the contrary; an assumption of zero selective placement for all types of adopted pairs studied (including MZAs) despite direct empirical evidence to the contrary; and no allowance for environmental correlation for separated siblings in the 1978 analysis.

3. Their own assertions to the contrary, the Honolulu model and specifications as applied to the study of collateral pairs do show certain important sensitivities to plausible perturbations in both their data and their model. An illustrative analysis using iterative weighted least squares showed that for data changes: (a) h^2, c^2, and q^2 all decrease as the observed r for siblings or DZ twins increases; and (b) c^2 (child environmentability) increases as the observed r for separated MZ twins decreases. Regarding changes in the model, when values for the collateral-kin environmental correlations (u_i) were introduced, it was found that as the differences among the u_i across collateral pairs increase, adult heritability, q^2, remains unchanged but child heritability, h^2, decreases markedly while child environmentability, c^2, increases markedly. As the u_i became more discrepant for the 1978 data set, the fit of the modified model actually improved beyond that of the Honolulu specification. There is thus reason to be skeptical of the researcher's claim that their (not wholly explicit) assumption of equal within-pair environmental correlation for collateral pairs consistently gives a good fit. Yet Rao, Morton, and Yee have confidently asserted that "the biological and cultural factors in the inheritance of IQ are resolved" (Rao et al., 1978, p. 447) and even (in two separate publications) that "the 'nature-nurture' controversy was partly an ideological confusion of individuals and populations, partly a methodological problem in distinguishing cultural and biological causes of family resemblance. As far as that problem has been formulated, it is solved" (Rao and Morton, 1978, p. 172; Morton and Rao, 1978, p. 37). On the other hand Morton in 1972 argued that "no intelligent skeptic would be converted by a heritability estimate, which a geneticist finds unconvincing" (Morton, 1972, p. 257).

The Honolulu Data Sets

There are some specific problems with the Honolulu group's 1976 (Table 5.1) and 1978 (Table 5.2) data sets, only some of which are touched on here. The interested reader should consult Goldberger (1978b) for a full treatment. (Goldberger's discussions are based in part on an analysis by L. Kamin, personal communication, 1979.)

For the 1976 data set, Rao et al. (1976) used data from the Jencks et al. (1972) compilation plus various correlations from the massive Burks (1928) study of adopted children. None of the studies used to compile the 1976 data set was done after 1940. Goldberger (1978b) also notes the considerable heterogeneity around the weighted means for the observed r values appearing in Table 5.1 for at least four kinship categories, sufficient to throw into question any calculation of means. For the IQ correlation of spouses (their kin number 11, designated FMTXY), they use the Jencks et al. mean of .50 based on 887 pairs, as shown in Table 5.1. But this mean is based on seven studies with observed correlations ranging from .40 (141 pairs from the Willoughby, 1928, study) to .74 (51 pairs from Outhit, 1933). For the correlation between the IQs of adoptive parent and child (kin number 9, designated OFPXY), the mean r of .23 (based on 1,181 pairs) is obtained from six studies ranging in this correlation from .07 (178 pairs from Burks, 1928) to .37 (180 pairs from Freeman et al., 1928). Their mean r of .52 (2,001) for siblings (kin 3, SSTXY) is (as noted in Chapter 2) based on heterogeneous correlations and shows, for example, an r of .42 (for Willoughby's, 1928, "about 280" pairs, used by Jencks et al. although uncited by them) and .67 (63 pairs from Outhit, 1933). In arriving at their averaged correlation for adopted-adopted pairs (kin 4) of .23 (from Burks's 21 pairs), they discarded the high Skodak (1950) correlation of .65 ($N = 41$) as well as the Freeman et al. (1928) correlation of .40 (based on 93 pairs which, it will be recalled, includes the 21 Terman cases which are the same as the Burks pairs). This of course considerably lowers their final value for adopted-adopted pairs, thus somewhat lowering the solved value for child environmentability (c^2; see equation [4], Table 5.1) as well as bringing this observed correlation considerably closer to the predicted correlation ($\hat{r}$) from their model (which, incidentally, comes out to .092), thus lessening their overall χ^2 value, thus improving the fit of their model. As pointed out in Chapter 2 Jencks et al. themselves averaged together four studies in this category; they arrived at a weighted r of .42 ($N = 165$). Despite this, Rao et al. use the same sources that Jencks et al. use in order to get the correlation for adopted-natural pairs (Freeman et al., 1928; Leahy, 1935; and Skodak, 1950).

Rao et al. (1976), unlike Jencks et al. (1972), explicitly assume zero selective placement, and thus zero correlation between the environment of the natural parents of an adopted-away offspring and that of the adopting parents. They do this despite the following available correlations (cited in Jencks et al.) which directly contradict their assumption: the correlation between adoptive parent's education and natural parent's education has been found to be .31 ($N = 124$) from Leahy (1935); .29 ($N = 836$) also from Leahy; .25 ($N = 94$) from Leahy; and .27 ($N = 100$) from Skodak and Skeels (1949). Scarr and Weinberg (1976) find a correlation of .22, which is also ignored by Rao et al.

For the 1978 16-kin data set, Rao and Morton (1978) use the kinships and studies from their 1976 data set plus additional correlations pertaining to environmental indexes (SES) from Duncan et al. (1972), Duncan and Featherman (1973), and Scarr and Weinberg (1977). In general, the rather marked heterogeneity persists in certain kinship categories (especially kin categories 3 [siblings], 4 [adopted-adopted pairs], 9 [IQs of adoptive parent and child], and 11 [IQs of spouses]), despite the addition of studies to some categories (for example, three studies to kin category 3; one to kin 4; and two to kin 11). Furthermore, much new data are excluded from kin categories for their 1978 compilation. Goldberger (1978, p. 91) notes for example that 14 usable new correlations are yielded by the Scarr and Weinberg (1978) study (which was available in manuscript form when Rao and Morton, 1978, compiled their data in the fall of 1976). While Rao and Morton use an average r of .570 ($N = 1,272$) for kin 8 (the correlation of parent's IQ with their own child's SES index), all four values from Scarr and Weinberg (1978) for this kin category are considerably lower: .13 ($N = 237$); .19 ($N = 150$); .37 ($N = 237$); and .40 ($N = 150$). And while they use an r of .511 for kin 11 (the correlation of the IQs of spouses), the Scarr and Weinberg data show correlations of .24 ($N = 120$) and .31 ($N = 103$). Finally, Goldberger (1978b, p. 92) notes that four new kinship categories (thus four new equations) might have been added to the Rao and Morton analysis since studies they did use contained correlations for these unused kinships: the IQ of a child with his subsequent occupation ($r = .36$, $n = 4,386$); the IQ of a wife with father-in-law's occupation ($r = .19$, $n = 4,386$); the IQ of a parent as child with the IQ of own child ($r = .44$; $n = 2,032$); and the IQ correlation of half-siblings ($r = .44$, $n = 50$, after Nichols as cited in Loehlin et al., 1975, p. 119).

A NOTE ON THE BIRMINGHAM MODELS

While the Honolulu research follows in the path analysis tradition of Wright, the Birmingham research (so called after J. L. Jinks and L. J. Eaves of the department of genetics at the University of Birmingham, England, and D. W. Fulker formerly of the University of Birmingham) follows in the classical biometrical-genetic tradition of R. A. Fisher. This research is briefly mentioned here as a contrast to the Honolulu research; but to do the topic justice the reader should consult Goldberger (1978a, 1978b, pp. 1–37, pp. A1–A15) and Loehlin (1978).

Like the Honolulu group, the Birmingham group employs least squares fitting for an overidentified set of equations for different kinships. The essential differences between the Birmingham models (especially those of Jinks and Eaves, 1974; Eaves, 1975; and Fulker, 1975) and the Honolulu models (Rao et al., 1976, 1978; Rao and Morton, 1978), are as follows:

1. The Birmingham models greatly deemphasize the role of gene-environment correlation, generally assuming it to be zero. (Jinks and Eaves, 1974, p. 289 include a parameter for gene-environment correlation; it came out to $-.31$ and failed "significantly" to improve the fit of their model to data from Burt and from Jencks et al.; nor according to them was this value significantly different from zero.)

2. The Birmingham models have no parameter analogous to Honolulu's f, for the effect of parental environment on their own offspring's environment.

3. The Birmingham models have no parameter analogous to Honolulu's x, for the effect of parental IQ on their own offspring's environment.

4. The Birmingham models have no provision for some empirical indicator (index) of aspects of social environment, which the Honolulu models do (their I_F, I_M, and I_C in Figure 5.2).

In other words, the Honolulu models are considerably more specific about environmental effects ("cultural transmission").

5. The Honolulu models always assume that the parent's contribution of genes to two (or more) offspring is such that the path from G_M or G_F to G_C is always $\frac{1}{2}$. Thus, assuming no assortative mating such that $m = 0$ (a Honolulu assumption), $r_{G1G2} = (\frac{1}{2})^2 + (\frac{1}{2})^2 = \frac{1}{2}$ for siblings and DZ twins. This is the expectation assuming random mating and no genetic dominance. But the Bir-

mingham specification is more complex; it incorporates provisions not only for assortative mating and dominance within this definitional formulation for r_{G1G2}, but allows r_{G1G2} to vary according to the amount of environment shared by parent and offspring (although this latter quantity is assumed in analysis to be zero).[4] In general, the Birmingham researchers (especially Jinks and Eaves, 1974) greatly emphasize the role of dominance in the determination of phenotypes such as IQ, while the Honolulu researchers not only consider the notion of dominance in the case of IQ to be far fetched ("The notion of dominance deviations for polygenes seems farfetched"; Morton, 1974, p. 320) but indeed preposterous: "Today the geneticist who fails to differentiate between environment common to children and parents and ascribes any excess of sib correlation over parent-offspring correlation to dominance must defend his integrity and intelligence" (Rao et al., 1976, p. 241). Incidentally, after reading this, one again wonders about the cross-validation and the "consensus" that Jensen, Eysenck, and Herrnstein often claim characterize the literature in behavioral genetics.

Loehlin's (1978) instructive model fitting of data from Jencks et al. (1972) introduces a separate path coefficient for dominance (d) and finds the variance in IQ presumably explained by dominance (d^2) to be a reasonably high .28 (Loehlin, 1978, p. 427). His narrow heritability coefficient comes out to about .38; thus his broad heritability estimate is (.38 + .28) or about .66. Loehlin varied a number of assumptions, for example a correlation for selective placement (set at zero versus .15); gene-environment correlation (set at zero versus "various values"); setting either or both of the Honolulu paths f and x at zero versus nonzero; and allowing for differences in assortative mating. He finds that varying these assumptions makes little difference in the estimates of parameters such as h^2, c^2, and d^2. He does not vary two or more environmental correlations (the u_i) simultaneously across collateral kinships. (He does, however, vary one single u_i representing the correlation between the parents' childhood environments and finds little effect on the resulting parameter estimates.) He relies heavily on the Jencks et al. data compilation and excludes from his analysis all the correlations, and thus equations, involving environmental indexes.

Goldberger (1978b, pp. 13–37) reanalyzes Jinks and Eaves (1974) and Eaves (1975), noting that the Honolulu and Birmingham groups vigorously disagree in their heritability estimates. Rao and Morton (1978) showed a child heritability of .69 while Jinks and Eaves (1974) show it as .34. Rao and Morton's estimate of adult heritability is around .30; for Jinks and Eaves it is .68—the same as *child* heritability for Rao and Morton. (I again question the cross-validation and consensus claimed by Jensen, Herrnstein, and Eysenck.)

Goldberger finds among other things that: (1) Using a Birmingham parameter e, for environment common to raised-together siblings, and correctly constraining one of their parameters (A) according to the logic of R. A. Fisher's original formulation (the Birmingham group did not do so originally, but incorrectly left A as a free parameter), solving their 9-kin data set on data from Jencks et al. shows an extremely poor fit ($\chi^2 = 120.74$ for 4 parameters, thus 5 degrees of freedom). (2) In general, their method of estimating broad heritability is highly sensitive to only a few observed correlations (MZTs, MZAs, and adoptive parent-child correlations) and also to relatively small sample sizes. (3) They pool from Jencks et al. the adopted-adopted pairs with the adopted-natural pairs, thus ignoring the (great) differences in these two sets of observed correlations. (4) Their analysis of British data uses only Cyril Burt's data. (5) Their heavy reliance on the Jencks et al. compilation thus ignores the heterogeneity underlying several kin categories (especially DZTs, SIBTs, and as already noted the adopted-adopted and adopted-natural pairs).

Chapter 6

Summary and Conclusions

Earlier the IQ game was defined as the use of assumptions, many of which are implausible as well as arbitrary, to arrive at a numerical value for the genetic heritability of human IQ scores on the grounds that no heritability calculations could be made without benefit of such assumptions.[1] These assumptions and the pertinent data, procedures, and calculations have been looked at in detail. What conclusions can be drawn about the genetic heritability of IQ?

1. *The genetic heritability of human IQ scores within a given population is not a reliably estimable quantity.* The actual magnitude of the heritability coefficient is not at the present time estimable within a reasonably narrow range and with any degree of confidence. The Honolulu finding of child heritability of .69 and adult heritability of .30 juxtaposed against the Birmingham finding of adult heritability of .68, coming as they do from two of the most sophisticated bodies of research, in no way increase one's confidence in the reliability of heritability estimation. Those who seem to cover the moderate middle ground with heritability estimates in the neighborhood of .45, like Jencks et al., neither increase one's confidence nor decrease one's skepticism. In any analysis, the number of unknowns, including the heritability coefficient itself, is always in excess of the empirical information necessary to solve for those unknowns, which causes many researchers to make implausible assumptions simply so they can crank out some kind of numerical estimate of IQ heritability. It has been shown in several contexts that any equation for the IQ correlation for a given kin pair will contain a number of unknowns, and that several new unknowns must be introduced with each new equation. Whether one is in effect attempting to solve a single equation, or

equations and thus kinships two at a time, or many equations at a time through least squares procedures, the number of parameters to be solved is consistently and inevitably greater than the number of equations. This is perhaps the single major difficulty in trying to do what the literature on IQ heritability is trying to do. One can conclude with confidence only that the heritability of IQ is somewhere between zero and 100 percent.

2. *There is no hard and convincing evidence that the heritability of IQ is anywhere near substantial.* Certain researchers such as Jensen, Herrnstein, and Eysenck have argued that IQ heritability is indeed "substantial," at around 80 percent or even higher. Others (such as the Honolulu group when talking of child heritability but not adult heritability) conclude in favor of only slightly lower heritabilities. Still others (for example Jencks et al. and Scarr and Weinberg) have concluded in favor of IQ heritability in the moderate to high moderate range. But even granting many of the assumptions used, concluding in favor of low IQ heritability—even for children—is most plausible. That does not mean that IQ heritability is definitely low. It means that concluding in favor of low heritability is extremely plausible. There is certainly little evidence that would permit one to speculate about any genes "for" IQ. Consider in these connections the following major points:

a. The heritability estimates made in Chapter 2 via Jensen's equation using both the Jencks et al. compilation for collateral kin and this compilation plus various omitted studies ranged from virtually perfect to near zero. The more studies were included, the wider the range, which is totally uniformative—heritability could be very high or it could be near zero.

b. Of the 22 pairs of identical twins who were raised in reasonably separated ("minimally similar") social environments, whose averaged IQ correlation came out to .43, fully *half* had been reunited or raised in related families, or both, thus inflating even this low correlation. In addition, the correlation is probably inflated by other sources as well, such as sample bias, same-examiner bias, and age-IQ correlation (especially for the Newman et al. and Juel-Nielsen studies). Slight inflation also arose from Jensen's combining of Juel-Nielsen's time$_1$ and time$_2$ scores and transforming Shields's scores. Furthermore, the more separated among the "separated" pairs revealed an absolute intrapair IQ difference that approached the theoretically expected IQ difference (1.13σ, or about 17 IQ points) for pairs of *genetically unrelated randomly* chosen individuals from a general (nontwin) population.

c. When more realistic values for the MZA correlation of .40, .30, and .20 were used in the Jencks et al. path model for separated twins, h^2 ranged from .34 to slightly less than zero. Yet researchers continue to use inflated values for the separated twin correlation: Jencks et al. use .75, Fulker (1975) uses .95 (which he later admitted, in a personal communication, was "rather a suspect statistic"); Rao et al. (1976) and Rao and Morton (1978) use .69; Jinks and Eaves (1974) use .75 (from Jencks et al.) and .87 (from Burt); and Loehlin (1978) uses .68. These values all lump together separated with not-so-separated twins; truer values—as *input* data for an analysis—are much lower. Jensen has insisted that the fact that separated MZ twins are supposed to correlate more highly in their IQs "than DZ twins raised together *leaves really no doubt* of the heritability of IQ" (1975, p. 177, italics added; cf. 1978). Since the correlation for actually separated MZ twins is the same or less—probably considerably less—than that for DZ twins raised together, by Jensen's own reasoning, there should be considerable doubt about the heritability of IQ.

d. In Chapter 4 it was shown that in pairwise comparisons among different collateral kinships (MZTs, DZTs, SIBTs, adopted-natural pairs, and adopted-adopted pairs), as the unmeasured environmental difference Δr_{E1E2} approaches the measured difference in IQ correlation between two different kin (Δr_{Y1Y2}) across differing genetic correlations (Δr_{G1G2}), artificial inflation of the bogus heritability estimate is so marked that actual heritability (the "true" h^2) approaches zero. When one compares MZTs (who have non-zero gene-environment correlation such that $r_{GE(MZT)} > 0$) to adopted-adopted pairs (UNT-AA, who necessarily have zero gene-environment correlation, so that $r_{GE(UNT-AA)} = 0$), then any pairwise heritability estimate inadvertently subsumes the whole of the MZT gene-environment covariance and is thus artificially inflated by it. This affects not only Jensen-style pairwise heritability estimates, but the Jencks et al. application of the Jensen equation and any least squares solutions involving MZT versus UNT-AA comparisons.

e. In the discussion of Jensen's (1975, 1976) three-equation solution (Chapter 5), h_2 values as low as .15 (and as high as .72) resulted from the use of plausible values of the environmental correla-

tion. This hardly tells us anything useful about the relative
influence of genes and environment on IQ scores.

f. In the reanalysis of the Honolulu 1976 and 1978 data sets, one
parameter (u_i) was varied across collateral kinships, retaining all
other Honolulu assumptions and specifications. Both child heritability (which decreased markedly) and child environmentability
(which increased markedly), were sensitive to this modification.
Some of the child heritability values were as low as .30, although
according to the Honolulu results child heritability is supposed
to be "substantial." This does not mean that child heritability is
definitely low, but rather that there is evidence in favor of low
child heritability in the very data and approaches that have been
used as evidence of higher heritability values.

3. *Although low IQ heritability is a plausible conclusion, the idea
that the heritability of IQ could be literally zero is not entirely convincing.* In this respect, I differ somewhat from Kamin who concluded
after an extraordinarily detailed analysis of data quality, that "there
exist no data which should lead a prudent man to accept the hypothesis that IQ test scores are in any degree heritable" (1974, p. 1). I have
seen no methods and data that would lead a prudent person to reject
a hypothesis in favor of low, or even very low, IQ heritability. Zero
heritability remains of course a possibility, but does not seem plausible or likely for at least four reasons: First, while instances of zero or
near-zero heritability (and even of negative heritability) were observed in certain of the reanalyses, these seemed more the exception
than the rule. Second, the finding (for example by Skodak, 1950, as
used by Morton and Rao, 1978; Scarr and Weinberg, 1977), that the
correlation between the IQ (and education) of the natural parents
and their adopted-away child is often reasonably high, sometimes
even higher than the IQ correlation for adopting parents and their
adopted children, though it rests on a small data base, is striking. (On
the other hand, at least one study, that by Snygg, 1938, cited by Kamin,
1974, p. 112, found a correlation of only .13 between the natural
mother's IQ and that of the adopted-away child, and this could easily
be explained by selective placement. Kamin also makes the interesting observation that the Snygg study has simply never been cited in
the literature on adopted children; it has become a forgotten study.
Nor is it cited in the research reviewed here, including the Honolulu
publications, which *do* use this particular kinship category in analysis.) Third, there seems to be little disagreement in the literature, in-

cluding the critical literature, that certain IQ-related phenomena at the extremes of the IQ distribution, especially certain forms of severe retardation (for example, Down's syndrome or Phenylketonuria) as well as perhaps certain forms of giftedness, have at least in significant part a genetic basis. If such extremes are averaged in with all other data, it is difficult to imagine how the *average* IQ heritability in a population would be literally zero. Fourth, according to Morton, it has been "the experience of biometrical genetics that a trait heritable at its extremes has never been found to have zero heritability within the normal range" (1972, p. 257), an observation that does not, of course, rule out low heritability.

4. *There is absolutely no known relationship between within-group heritability (for either whites or blacks) and the heritability of the average black-white IQ difference. Furthermore, recent evidence shows little, if any, connection between biological (ancestral; blood-group) measures of race and IQ score.* In Chapter 2 it was shown *even if* one had reliable estimates of IQ heritability for both blacks and whites, there would be no way to make even a vague estimate of the heritability of a between-group average IQ difference from within-group heritability. Geneticists have long known this; it is Jensen, Eysenck, and Shockley who have not. Furthermore, I have seen no convincing evidence that the within-group IQ heritability for whites is anywhere near "substantial." There are only four known studies of within-group IQ heritability for blacks, and their results are not only inconclusive in themselves, but suffer the same problems as studies of IQ heritability among whites—which has been the subject of most of this book. Finally, there is little evidence for any relationship between IQ and race as measured by biological criteria (such as blood group). In fact there are some bits of evidence, which are only suggestive, that higher odds of African ancestry are associated with higher IQ performance. I do not seem to read about that set of findings in the writing of Jensen or Eysenck, and certainly not in those of Shockley.

This book has been primarily about the kinds of assumptions that are used in heritability estimation, what is wrong with those assumptions, and how they affect the results. I would like now to summarize the most important and most implausible assumptions. For each assumption listed here, I review some evidence that runs directly counter to it:

1. *Equality of environmental similarity (correlation) across kinships.* Direct evidence against this assumption exists. For example, there are

consistent differences in IQ correlation between DZ twins and ordinary siblings (who are genetically the same as DZ twins). Same-sex
DZ twins correlate more highly in IQ than opposite-sex DZ twins, a
clear nongenetic effect. MZs are treated more similarly in the environment than are DZs. If one prefers to place this source of IQ variance in
gene-environment covariance rather than in the environmental similarity quantity, the resulting inflation of the heritability estimate (via
pairwise comparison of kin) can be even more marked. I noted in several contexts that varying the environmental similarity correlation
(r_{E1E2}; that is, u_j) across kinships changes the heritability estimate markedly; placing MZ-DZ environmental differences into a residual term
(as done by Jencks et al.) is not a substitute for varying the intrapair
environmental term. (Other researchers, for example, Hogarth, 1974;
Jensen, 1975, 1976; and Goldberger, 1977, who clearly demonstrates
the sensitivity of the h^2 estimate to this quantity, have varied r_{E1E2} across
MZ-DZ comparisons, but here the analysis was extended beyond MZ-
DZ comparisons.) In fact one may say as a general conclusion that the
failure to develop a way of permitting the environmental correlation
quantity to vary across analyzed kinships (both collateral and direct
line) is a major weakness, perhaps the major weakness, of present-day
methods for estimating the heritability of IQ.

    2. *Zero environmental similarity for separated kin, including that of
natural and adoptive parents.* This assumption is a familiar one in the
study of separated MZ twins (with Jencks et al. the exception), separated siblings, and any other separated collateral kin. Direct evidence
against this assumption exists in the case of the separated twins. This
assumption also operates if the researcher (like the Honolulu group)
assumes no selective placement of adopted children: This forces the
correlation between the environments of natural and adopting parents, as well as that between the environments of natural parents and
their adopted-away children, to be zero. Some researchers (notably
Jencks et al., 1972; Scarr and Weinberg, 1977; Loehlin, 1978) do allow
such correlations to be nonzero. Gene-environment correlation is
generally assumed to be zero for separated kin, especially (implausibly) for separated MZ twins.

    3. *Equality of gene-environment correlation across kinships.* There is
in fact reasonably direct evidence to the contrary: One can be confident that gene-environment correlation for identical twins raised
together is greater than zero; their genetic identity results in high environmental similarity. Gene-environment correlation for adopted-

adopted pairs is probably zero or nearly so. Thus, at least some kin differ from others in the amount of environment correlation. I suspect a fortiori that other compared kin might also. The assumption that cross-pair gene-environment correlations (r_{G1E2} and r_{G2E1}) are equal for any given kinship seems dubious.

4. *Recursiveness.* This assumption—used in actual analysis, most often implicitly, by every researcher considered here—leads necessarily to such nonsensical inferences as these: A person's IQ score does not affect how he is treated in his own environment; the IQ scores of the members of any given pair have no direct causal effect on each other; one's socioeconomic status does not directly affect one's own IQ (the Honolulu models).

5. *Residual noncorrelation.* Every formulation considered here allows for no correlation whatever between any residual and any other variable (other than its own) in the entire model. For Jencks et al. and the Honolulu researchers this assumption is explicit. For Jensen it is sometimes explicit (for example, 1976) but most of the time implicit. For Eysenck and Herrnstein it is wholly implicit; in fact neither gives any evidence of awareness of the assumption at all. Such an assumption requires the inference that one's residual (nonfamily) environment has no effect whatever on the following: one's intrauterine environment, one's spouse's nonfamily environment, one's spouse's family environment, one's child's nonfamily environment, one's child's family environment, and one's child's IQ score.

6. *That the samples used are representative of their respective populations.* There are many reasons to suspect that the samples in the separated-twin studies were biased toward underrepresentation of environmentally dissimilar twins, and there is reason to be suspicious of the samples in most studies of other kin categories as well. For example, the samples are rarely random, they frequently represent only a limited range of environments, and the sample sizes are frequently small.

7. *That measurement error is very low, random, and appropriately eliminated by the attenuation correction.* I strongly question whether currently used IQ tests, such as the Stanford-Binet, Wechsler, Raven's, and others, measure any unidimensional construct of cognitive functioning. I also question whether IQ tests represent interval scales and whether they are unbiased against minorities. The frequent age-to-IQ correlation suggests inadequate standardization. The popular attenuation correction procedure, used extensively by Jencks et al. and by

Jensen, requires an assumption of random measurement error. If any systematic bias exists (such as differential effects of variables such as race, sex, SES, and the like, upon two or more indicators of the same concept), then the attenuation correction is wholly inappropriate.

8. *That there are no prenatal or postnatal environmental asymmetries.* Although the issue of whether transfusion syndromes differentially affect twins' intrauterine birth weights is an open question, as is the question of whether there is a significant relationship between birth weight and IQ, it seems unwise arbitrarily to assume equality of intrauterine effects on IQ (if there are any) for twins. It seems even more arbitrary to assume that no asymmetries and role complementarities exist in the postnatal environment—that is, in social interaction, leadership roles, dominance relations, school, on the job, and so on, for MZ twins, DZ twins, siblings, or even unrelated persons raised together.

9. *Linearity* and *additivity.* I find these two assumptions dubious, but lack direct evidence: Any nonlinear relationship between genotype and phenotype would invalidate virtually all of the equations and procedures presently employed. The issue of gene-environment interaction versus additivity has been extensively discussed.

Other shortcomings, errors, and misrepresentations in portions of the IQ heritability literature have also been pointed out. These are listed next.

1. *Arithmetic errors and obfuscations.* Examples of simple out-and-out arithmetic mistakes are these: Jensen's finding "only one realistic" solution in his three-equation analysis; Kamin's (1978) observation that Munsinger (1977) had miscalculated Jensen's (already ill-defined) "difference correlation"; the errors appearing in certain tables in Jencks et al., reviewed in Chapter 2; and the miscalculations of some h^2 values in Scarr and Weinberg (1977). In the category of arithmetic obfuscations, the list is quite long: Jensen's artifactual and ill-conceived "difference correlation"; Jensen's shifting choice of significance levels (see "Gene-Environment Interaction" in Chapter 3); Jensen's use of pooled instead of averaged IQ correlations in his 1970 paper on separated twins (pooling inflates the correlation relative to averaging); Jensen's use of Juel-Nielsen's combined $time_1$ and $time_2$ IQ scores rather than only the $time_1$ scores; Jensen's (unspecified) transformation of Shields's separated twin scores; Fulker's (1975) use of an IQ correlation for separated MZ twins of .95; the nonreporting of negative heritability values in Scarr-Salapatek

(1971); the nonreporting of at least three negative heritability values in Scarr and Weinberg (1977); the two specification errors Loehlin et al. (1975) and Loehlin (1978) found in Jencks et al. (1972); and the specification error Goldberger (1978d) discovered in Rao et al. (1976).

Under the heading of near obfuscations one might include the following: the widespread and recent use of the simple-minded pairwise Jensen (1967) kinship equation in virtually all MZ-DZ comparisons, for example, Scarr and Weinberg (1977, 1978), the kinometrics research of Behrman and Taubman (1976) and Behrman et al. (1977), Jencks and Brown (1977), and Rao and Morton (1978) (because of the assumptions of zero gene-environment correlation and equal-environment correlation, which for Rao and Morton reduced five separate pairwise comparisons among collateral equations to a simple Jensen-style equation).

2. *Questionable inclusion or exclusion of studies and correlation values.* The following examples might be mentioned: the difficulties with the 1963 Erlenmeyer-Kimling and Jarvik figure, listed in Kamin (1974), reviewed in Chapter 2; and the Jencks et al. exclusion of studies and data on collateral kin, particularly as pertains to the possible overestimation of the sibling correlation r_{SIBT} in the population, and the possible underestimation of the DZ correlation r_{DZT}. The Honolulu researchers' reliance on the Jencks et al. compilation perpetuates many of its difficulties, such as the double counting of the 21 Terman cases, the guessed-at sample sizes (as with the Willoughby study), and the *great* heterogeneity of correlation values across studies within a kinship category. Furthermore, in several important cases data reported by Jencks et al. (such as the selective placement correlations between the educations of adoptive and natural mothers) were ignored by the Honolulu group.

3. *Misquoting, misciting, and incorrect transcription.* Jensen's misquotations from the separated twin studies involved sample sizes, correlation sizes, age at separation, the extent of environmental separation, and other matters (see Chapters 2 and 3). Herrnstein also misquotes the literature which he claims to have searched "from one end to the other" (1973, p. 9), as does Eysenck (see Chapter 1; for a more complete treatment of this aspect of Herrnstein, see Goldberger, 1976). Vandenberg (1971) incorrectly transcribed data on separated twins from Newman et al. and Juel-Nielsen. Jencks et al. miscopied some of Burt's correlation values, excluded the Willoughby study from the appropriate table, reversed two sample sizes for two correlation values for married pairs; and reversed the Conrad and Jones sib-

ling correlation with the parent-offspring correlation among other errors.

This entire book has been a critical evaluation of important studies in the heritability literature. This criticism has been motivated by a sincere concern with the accurate assessment of the causes of human variation in IQ scores. If IQ heritability within a given population is to be reliably investigated, the problems listed below must first be solved. Whether they *can* be solved in the next few years is, of course, another question.

1. A way must be found or devised to permit *variability of a quantity for intrapair environmental similarity across collateral and direct line kinships and adoptions*. The analysis in Chapter 5 is only suggestive in this respect. Procedures for the independent empirical measurement of this quantity are needed. This will be difficult. More attention should be paid to the quantification of treatment effects—the systematic comparison of kinships of equal biological similarity that differ in IQ correlation. More attention must be given to the DZT-SIBT comparison. Obtaining the correlation between SES characteristics of natural and adopting parents in the study of separated pairs is another approach. Measurement of the correlation between parent's SES and adult offspring's SES is another (for direct line pairs). Clearly, environment common to parent-child pairs must be distinguished from environment common to collateral pairs. The use of SES indicators as measures of environmental similarity or common environment has already been questioned. The issue of independent empirical measurement of any environmental correlation difference between kinships still remains and is one of the most important issues in the entire field of genetic causes of IQ variation.

2. A way must be found to permit *variability of gene-environment correlation* across collateral and direct line kin and adoptions. This includes especially MZT-DZT-SIBT-UNT comparisons. (It is necessary though certainly not sufficient to distinguish only adopted-adopted pairs and adopted-natural pairs in this regard.) Perhaps environmental differences between MZTs and DZTs, for example, should be assigned to gene-environment correlation rather than to environmental correlation, as argued above. The point is that it must be assigned to one or the other. (Assigning it to the residual term as do Jencks et al. for MZ and DZ comparisons is unwise, for at least two reasons: First, it throws this source in with all the other sources of variance in the residual

term. Second, it thus does not permit one to assess the amount of IQ variance resulting from the difference.) Gene-environment correlation is very likely not equal across certain kinship comparisons (recall the MZT–adopted-natural pair–adopted-adopted pair comparison). Yet no research considered here has permitted differences in gene-environment correlation among other kinships such as MZTs versus DZTs versus SIBTs, or even for general (unpaired) individuals in a population versus everyone else, and for other comparisons.

3. *A construct (concept?) for which IQ score is an "index" must be included in one's model.* In a sense, "IQ" is not the phenotype; perhaps the phenotype is "intelligence" or some such concept. (Yet phenotype is always defined as a directly observable variable, that is, an indicator or index itself.) It is extremely clear that IQ is intended in all formulations in the IQ heritability literature, whether explicitly or not, to serve as an indicator or index of some concept (intelligence; smartness; brightness; cognitive functioning; whatever). Otherwise, one runs the risk of being accused of repeating Boring's (1923) stark extreme operationism—that intelligence is what an intelligence test measures. And no one I have encountered wants to be accused of that. What this means is that some concept representing "intelligence" or whatever must be built into one's model. (Simply blindly correcting for attenuation is obviously no substitute.)

4. *Measurement models might be developed to include multidimensional indexes* of "intelligence." *All* models (including path models) but one have only one dependent variable (IQ). (The exception is the kinometric approach, but there the *dependent* variables are SES variables and not IQ.) This is true even if different subtypes of test performance (for example, verbal versus quantitative) are measured; only one is used at a time in a given model or heritability calculation. Not only should a concept (or construct) "for" this single indicator be included, but an allowance should be made for two or more indicators for the same concept in the *same* model, and for systematic measurement biases as well (see Chapter 4). Certainly, the continued routine application of the attenuation correction to any and all IQ correlations might profitably be abandoned.

5. *The ill-defined, methodologically inadequate Jensen-like pairwise equation* for the calculation of the heritability coefficient should be abandoned. This would of course mean eliminating h^2 estimates based on: the difference in the IQ correlation for identical and fraternal twins; the previously cited kinometric research; formulations such

as that in Jencks and Brown (1977); and the Jencks et al. effort to apply the Jensen equation to five collateral kin. The heritability estimates of Scarr-Salapatek (1971, 1974) and Scarr and Weinberg (1977, 1978) would also be eliminated.

6. The use of *nonrecursive models* would help in alleviating a number of the problems related to the effects of one's IQ upon one's environment, the reciprocal effects of the IQs of collateral pairs on one another, and any other reciprocal effect.

7. The conceptually most attractive collateral kin designs involve separated identical twins (maximal genetic similarity with minimal environmental similarity), and unrelated adopted-adopted pairs raised together (minimal genetic similarity with maximal environmental similarity). The only study of separated identical twins in the United States is the small, old, biased-sample Newman et al. study. The only British study (excluding Burt's discredited work) is the Shields study. Obviously, a well-designed large study of MZ twins separated at birth and randomly allocated over a wide range of environments would be intriguing. A study under way at the Department of Psychology, University of Minnesota, under the direction of Professors T. Bouchard, I. Gottesman, L. Heston, and others, shows some promise in this regard (Bouchard, personal communication, 1980). (Such a study would not rule out the kind of analysis done in Chapter 3, where the amount of environmental similarity after separation was explicitly taken into account.) Perhaps the converse design, a well-planned large study of randomly placed (or nearly so) adopted-adopted pairs would be feasible (the Scarr and Weinberg research approaches this to some extent). But even approximate random allocation of humans into homes with no selective placement seems just as impossible on practical as on ethical grounds, leaving little reason for optimism about either type of study.

In the final analysis, what we are left with is a mass of faulty methods and data, which do not permit one to conclude in favor of a significant genetic effect on IQ score. To the extent that the literature "tests" a hypothesis that human IQ is substantially genetically caused, the hypothesis must be resoundingly rejected. Given the very real policy implications not only for minorities but for everyone else as well, we must have considerably better evidence before concluding that substantial IQ heritability exists. There is as yet no compelling reason to postulate the existence of any genotype for IQ score; certainly no reason arises from the calculation of the heritability coefficient. I am reminded again of Morton's comment, after noting the skeptic's resistance to the idea that certain human behavioral traits such as IQ are to any important degree genetically caused: "In any case, no intelligent skeptic would be converted by a heritability estimate, *which a geneticist finds unconvincing*" (Morton, 1972, p. 257, italics added). I am not quite convinced, either.

The Cognitive Style Challenge to IQ

Daniel R. Vasgird
Department of Sociology
University of California, Berkeley

Significantly, IQ has never been a very satisfactory predictor of creativity, although it has been shown to be moderately predictive of socioeconomic success in American society (cf. Jencks et al., 1972). The test reflects an orientation to reality that complements the immediate needs of technology and industry. In this respect, IQ is an indicator of certain socially subjective priorities. It has not been predictive of certain cognitive activities that have been judged by some to be important in the long-run development of a culture, for example, creativity and empathy. And it is not simply noncognitive factors that lift the individual into these realms, contrary to Herrnstein's speculation (1973). Herman Witkin, Robert Ornstein, Roger Sperry, Rosalie Cohen, and others have been doing research on cognitive styles that indicates two kinds of thought—the analytic and the integrative. Many of these researchers believe that IQ tends to measure only the analytic. What follows is an overview of how the interpretation of cognitive style development has evolved over the last quarter century.

This evolution has passed through three distinct phases. A social learning interpretation was adopted by the early researchers into cognitive styles, who had a distinct preference for the analytic style. The second phase, which began in the late sixties, maintained a social learning explanation, but was characterized by an equal regard for both styles. Recently, however, a neurophysiological hypothesis has evolved that is theoretically similar to the interactionist school of general cognitive development.

WHAT ARE COGNITIVE STYLES?

Cohen (1969) defines cognitive styles as integrated rule sets for the selection and organization of sense data. This entails a broad conception of cognition and implies that within each person we find a unique qualitative composition of the array of cognitive processes such as sensation, perception, and imagery, that is, a unique integrated rule set. For example, some individuals habitually analyze perceptual stimuli point by point and have difficulty in using global scanning techniques that can be beneficial in discerning extended interrelationships (Bogen et al., 1972; McClelland, 1967; Witkin and Goodenough, 1976). Furthermore, people can also differ in their retention processes, some stressing visual imagery and others verbal (Luria, 1976). What method is used appears to depend essentially on what people have been exposed to and found beneficial in the past, although some researchers, such as J. Levy at the University of Chicago (reported in Cromie, 1978), argue that there are genetic differences in brain organization between males and females and that these have a strong effect on cognitive behavior.

Cognitive style is therefore a synthesizing conceptual tool for summarizing a person's basic cognitive approach to all manner of sensory stimuli. Kogan embodies this conceptualization when he defines cognitive styles as "individual variations in *modes* of perceiving, remembering, and thinking, or as distinctive ways of apprehending, storing, transforming, and utilizing information" (1973, p. 160, italics in original). Yet he also notes that "abilities concern level of skill—the more and less of performance—whereas cognitive styles give greater weight to the *manner* and *form* of cognition" (p. 160, italics in original).

This emphasis on manner and form serves to distinguish cognitive styles from the "intelligence" and even the "creativity" dimensions within which level of performance is used to differentiate individuals. Kogan, however, observes that even many of those who use the label "cognitive style" have based their assessments of cognitive style on accuracy versus inaccuracy of performance (for example, Witkin, et al., 1962). To be truly stylistic, matters of accuracy should be irrelevant, and, furthermore, "no value judgments [should be] placed upon the kinds of performances derived from the tasks employed to assess the style at issue" (Kogan, 1973, p. 161). (Significantly, Witkin and his associates, 1976, 1977, have recently espoused an opposite view. I discuss this shortly.)

Within the dimension of cognitive style itself, the stylistic types are generally theoretically and operationally dichotomized. The observation that human thought is dualistic is not a new one. Philosophers and artists have reflected on this dual nature. Pythagoras and Plato divided the mind into two parts, one partaking of reason, the other devoid of it. More recently, William James reflected on the "differential" and the "existential" aspects of the mind. Hobbes, Levi-Strauss, A. R. Luria, and others have made similar typifications of the dual mind. Freud divided mental activity into two areas, primary-process thinking and the more rational secondary-process thought, while Piaget extended the Freudian model by depicting the animistic mode of young childhood as being gradually supplanted by the scientific.

The *analytic cognitive style* is characterized by a formal mode of abstracting salient information from a stimulus or situation and by a stimulus-centered orientation to reality. It is parts-specific, meaning that parts of attributes of a given stimulus have meaning in themselves. The analytic style involves logic, and those who use it process information sequentially. The *integrative cognitive style,* on the other hand, requires a descriptive mode of abstraction. It is the global characteristics of a stimulus that have meaning to its users, and these only in reference to some total context. Information is processed more diffusely with the integrative style than with the analytic, and material is assimilated in a simultaneous rather than a linear fashion. There seems to be a spatial orientation to reality inherent in it.

THE SOCIAL LEARNING HYPOTHESIS

The social learning approach in the study of cognitive style development (Bandura, 1969) is by far the more traditional, as it is in the field of cognitive development. Its followers believe that the environment is the primary cause of distinctive cognitive styles. The variations within this group stem from the validity each attributes to different cognitive styles. One group places extreme value on the analytic while devaluing the integrative, and the other group values both cognitive styles. Within the former is the pioneering work of Witkin (1954; Witkin et al., 1962) and Kagan (1964). This formulation went uncontested until McClelland (1967) described an alternative to the analytic.

Witkin (1967) was a deeply committed social learning advocate who believed that the source of the processes that have been found to foster the development of an analytic cognitive style would be found in the school and other social settings. He laid the groundwork for much of the research being done today. And yet, until very recently, his scope was limited by his regard for the field-independent (analytic) cognitive style. Conceptually, his propositions made his work *appear* to deal with a cognitive ability of some sort, certainly not a stylistic variable as he stated. Empirically, there was no reason for him to downgrade the field-dependent (integrative) style without a direct measure for it. He conceptualized development as a progression from the integrative to the analytic style, using adjectives such as "limited" and "primitive" for the former (Witkin et al., 1962; Witkin, 1967). Individuals who used the integrative technique beyond their childhood years were depicted as cognitively stagnant and undeveloped.

Kagan et al. (1963) and Witkin helped demonstrate a link between perceptual process and cognitive style. Some of their most important conclusions related to age and sex: They found that children tend to become more analytic as they grow older, and that males are more consistently analytic than females. Though they recognized the integrative approach, they preferred to treat all cognition outside of the analytic category as nonanalytic.

EQUAL REGARD FOR THE INTEGRATIVE

The sixties had an effect on scholarly interest in the integrative style. McClelland (1967) and Cohen (1969) were instrumental in this transition. They maintained a social learning position, but at the same time gave much more credence to the validity of integrative thought. Certain anthropologists, however, on the basis of observations outside of our society had been considering the integrative since the early 1950s (see, for example, Lee, 1968). Cohen, McClelland, and others of their school are significant because they draw their conclusions from data derived within the United States.

In writing on sex differences, McClelland noted that it was easier to describe the kinds of situations and phenomena the male desires to confront in the world: the simple, the closed, and the direct. McClelland sees males as more analytic. "In contrast, women are interested in the complex, the open, and the undefined" (1967, p. 180). This gives rise to a desire for independent relationships in which the "whole picture" is important. He argues that highly integrative individuals are a valuable resource in society, one that has been underrated.

R. A. Cohen (1967, 1968, 1969; Cohen et al., 1968) has been engaged in research on cognitive styles for more than a decade at the University of Pittsburgh and was among the first to present both styles of thought as having equal validity. The units of analysis for her research are the family, the friendship group, and the school—the major socializing agents in society. Cohen, like Witkin and Kagan, determined that the conceptual environment of the family and friendship groups differed according to status grouping. The school, on the other hand, was found to be strictly an analytic environment in which "not only test criteria but also the overall ideology and learning environment of the school embody requirements for many social and psychological correlates of the analytic style" (Cohen, 1969, p. 830). In addition, she noted that many children demonstrate an integrative approach to organizing reality. When they are confronted with an inflexible school environment, culture conflict sets in, which she argued is not caused by what they do or do not know, but by conflicting styles of conceptual organization. A child's cognitive style is developed from the early socializing environment, and children of middle-class origin generally show an analytic approach while those of low-income and upper-class origin show an integrative approach. She went on to say that each cognitive style "could affect its carrier's ability to participate effectively in the alternate kind of group process or to deal directly with its cognitive requirements. In practice, it was found that children who had been socialized in shared-function (integrative) environments could not participate effectively in any aspect of the formal school environment even when native ability and information repertoires were adequate" (1969, p. 842).

Cohen finally concluded that American society discriminates against a valid and useful approach to selection and organization of information. She found that individuals adopt conceptual styles that are most beneficial to positive functioning in their particular social milieu.

Recently Witkin and his associates (Witkin and Goodenough, 1976; Witkin et al., 1977) have dramatically shifted to a position that fully recognizes both cognitive styles. They now use the constructs of field dependence and field independence (integrative and analytic, respectively), and individuals are pictured as *different* in how they move toward a goal, rather than as more or less "competent." The sheer abundance of research using Witkin's methods and concepts has made cognitive style nearly synonymous with his conceptualization. Now Witkin echoes McClelland and Cohen when he tells us: "There are indeed circumstances where a field-dependent or field-independent mode of functioning is *more adaptive* and the person with the mode that suits the given circumstances is *benefitted* by possessing it" (Witkin and Goodenough, 1976, p. 52, italics added).

THE INTERACTIONIST APPROACH

Witkin's early work coupled with a potentially explanatory neurophysiological phenomenon (such as the split-brain theory, discussed later) has all the explosive ingredients necessary to generate a conflict comparable to the so-called IQ controversy itself, perhaps even becoming subsumed within that controversy. But there is neither need nor justification to load the cognitive style dimension with value suppositions that would impede investigation into its nature, as individuals like Herrnstein (1973) have unwarrantedly done with IQ (see Vasgird, 1975).

Recent neurological literature, under the often distorting rubric of "split-brain theory" (for example, Bogen, 1969a, 1969b; Bogen and Bogen, 1969; Bogen and Gazzaniga, 1965; Bogen et al., 1972; Gazzaniga, 1970; Marsh et al., 1971; Kimura, 1963, 1964, 1973; Levy, 1970; Sperry, 1975; TenHouten, 1976, 1977) has given rise to what is called an *interactionist* explanation. This viewpoint states that the form of the mature individual is not a distinct product of the environment but rather a reflection of the interaction between the organism and the environment. Baldwin paraphrases the renowned interactionist, Jean Piaget, when he tells us that it is a "theory of human behavior which postulates a general cognitive mechanism as the initial step in the chain of events leading from the stimulus to the response" (1969, p. 328). This is what the split-brain theory predicts when combined with cognitive style hypotheses. It proclaims that human beings possess neural structures that work in tandem with social variables to give rise to the phenomenon of cognitive style differentiation.

While still new and tentative, the neurological research from which split-brain theory is drawn postulates that different cognitive processes originate in different hemispheres in the brain. The research attempts to demonstrate that there are neural structures in the brain that control certain cognitive abilities, and these structures can be hemispherically differentiated. Several theorists have suggested, on the basis of this research, that the brain's structural design is the progenitor for the potential to develop differential cognitive styles (Ornstein, 1973; Ramirez and Casteñeda, 1974; Samples, 1975; TenHouten, 1976). Ornstein offers a useful summary:

The cerebral cortex of the brain is divided into two hemispheres, joined by a large bundle of interconnecting fibers called the corpus callosum. The right side of the cortex primarily controls the left side of the body, and the left side of the cortex largely controls the right side of the body. The structure and function of these two "half-brains" influence the two modes of consciousness. The left hemisphere is predominantly involved with analytic thinking, especially language and logic. This hemisphere seems to process information sequentially, which is necessary for logical thought since logic depends on sequence and order.

The right hemisphere, by contrast, appears to be primarily responsible for our orientation in space, artistic talents, body awareness, and recognition of faces. It processes information more diffusely than the left hemisphere does, and integrates material in a simultaneous, rather than linear fashion (p. 87).

Neurological research on hemispheric differences has helped advance comprehension of integrative and analytic thinking. Consider the following abilities at which the nondominant right hemisphere appears to be superior to the dominant left: (1) part-to-whole assessment of visual or tactile stimuli (Nebes, 1972); (2) overall configuration recognition with no verbal processing requirement (Milner, 1971); (3) recognition and retention of nonsense and nonverbal stimuli (Kimura, 1963); (4) recognition and closure of a configuration of a stimulus from partial or incomplete information (Bogen et al., 1972; DeRenzi and Spinnler, 1966); (5) simultaneous use of all available information (Levy, 1970); (6) simultaneous assessment of events (Carmon and Nachson, 1971; Marsh et al., 1971); (7) use of musical stimuli (Kimura, 1964); (8) human face recognition (Benton and Van Allen, 1968, 1971; Yin, 1969, 1970); (9) drawing geometric expressions (Bogen, 1969a); (10) capturing and retaining the distribution and overall configuration of the components of information visually presented in three-dimensional space (Julesz, 1972).

The activities of the hemispheres are not exclusive of each other. The psychological literature emphasizes, too, that the spectrum of cognitive styles is a continuous mixture of two distinct modes. Yet one style tends to be emphasized over the other in development, and whichever of these is emphasized is an important determinant of the personality and overall psychological make-up of the individual.

In general, the interactionist cognitive style hypothesis based on split-brain data is supportive of Piaget's hypothesis that cognitive development entails innate cognitive structures in the human brain interacting with the environment to reach an equilibrium between potential and realized aptitudes. In this case, following Piaget's logic, the cognitively developed individual would be bicognitively fluent in terms of the analytic and integrative cognitive styles. As Witkin, who has also come to this conclusion, tells us: "The person who has access to both modes . . . has the potential for adapting to a wider array of circumstances, compared to the person who is fixed" (1976, p. 52).

CONCLUSION

Western society is decidedly fixed with its emphasis on the analytic cognitive style and its primary yardstick, the IQ test. Integrative qualities have been neglected until recently. This style of thinking combined with our contemporary obsession with the ranking of people caused by reputedly scarce resources and minimal opportunities to advance from socially inherited status, presents a bleak picture. But conceivably the IQ test might be revamped and broadened in scope, and the environment and tools of our education and media systems could be adapted to develop cognitive style flexibility.

Certainly cognitive style research could fall prey to the ranking obsession as IQ has been from its inception. Policy makers want rational tools to support disbursement of funds to groups and organizations; society needs "objective" means to allocate success to individuals. But we may well not need another ranking or stacking mechanism. Cognitive style research has the potential to account for individual cognitive variations in terms of means adopted to cope with the immediate environment rather than innate differences in "ability." It has the potential to assist in laying a foundation that promotes mutual respect and perhaps a more equitable distribution of resources and opportunities.

Appendix B

Coded Data on
Separated Identical
Twins

NEWMAN ET AL. (1937)

V_1	V_2	V_3	V_4	V_5	V_6	V_7
1	1	2	3	85	97	01
1	2	2	3	66	78	02
2	2	2	2	99	101	03
2	2	1	3	89	106	04
1	1	1	2	89	93	05
1	1	1	1	102	94	06
2	1	2	2	105	106	07
2	2	1	3	92	77	08
2	1	1	1	102	96	09
1	1	1	2	122	127	10
1	2	2	3	92	116	11
1	2	2	2	116	109	12
2	1	1	1	94	95	13
2	1	1	2	85	84	14
2	1	2	1	91	90	15
1	1	1	2	90	88	16
1	2	2	3	115	105	17
1	2	1	3	96	77	18
1	1	1	2	88	79	19

Note: V_1 indicates age at separation (1 = More than six months, 2 = Six months or less); V_2 indicates childhood reunion (1 = Reunited, 2 = Not reunited); V_3 indicates family relatedness (1 = Definitely related, 2 = Possibly related); V_4 indicates similarity in social environment (1 = Strong, 2 = Possible, 3 = Minimal); V_5 and V_6 indicate IQ scores for each pair member (untransformed [raw] scores are given for Shields's pairs); V_7 indicates case designation in original source (note that since Shields eliminated certain pairs [see Chapter 3], case designations here are not sequential).

SHIELDS (1962)

V_1	V_2	V_3	V_4	V_5	V_6	V_7
2	1	1	2	60	50	SM1
2	1	1	1	58	64	SM2
1	2	2	3	35	58	SM3
2	1	1	2	45	40	SM5
2	1	1	2	72	74	SM6
2	1	1	3	56	31	SM7
1	1	2	1	42	50	SM8
1	1	2	3	70	80	SM10
2	1	1	1	79	63	SM11
2	1	1	3	68	48	SM12
1	2	1	2	78	66	SM13
1	1	1	2	63	65	SM14
1	1	1	1	24	25	SM15
2	1	1	1	45	45	SF1
2	1	1	2	73	76	SF5
2	2	1	2	76	68	SF7
2	1	1	2	85	82	SF8
1	2	2	3	55	50	SF9
1	2	2	3	80	50	SF10
1	1	1	1	65	60	SF11
1	1	1	2	44	43	SF12
2	1	1	1	47	53	SF13
2	2	1	2	75	52	SF14
2	1	1	1	38	34	SF15
2	1	2	3	53	43	SF16
2	1	1	3	88	71	SF17
2	1	1	2	58	61	SF18
1	2	2	3	32	56	SF19
2	1	1	1	59	67	SF20
2	1	1	1	58	62	SF21
1	2	2	3	30	16	SF22
2	2	1	1	43	21	SF23
2	1	1	1	95	96	SF24
2	1	1	1	80	74	SF25
2	2	1	2	30	23	SF26
1	1	1	2	37	42	SF27
1	1	1	1	75	76	SF29

JUEL-NIELSEN (1965)[a]

V₁	V₂	V₃	V₄	V₅	V₆	V₇
1	2	2	3	119	121	01
				120	135	
1	2	2	3	99	103	02
				99	106	
2	2	1	2	108	97	03
				—	101	
1	1	1	1	91	100	04
				98	101	
1	2	2	2	111	117	05
				—	115	
1	2	1	2	105	97	06
				—	—	
2	1	1	2	100	94	07
				110	104	
2	1	1	3	91	98	08
				96	101	
2	1	1	1	104	103	09
				102	105	
1	1	1	3	125	111	10
				122	116	
1	1	1	2	111	117	11
				115	111	
1	2	2	3	99	112	12
				102	112	

[a] For Juel-Nielsen's IQ scores (V_5, V_6), the $time_1$ scores for each pair appear in the first row and the $time_2$ scores (where available) appear in the second row.

Notes

CHAPTER 1

1. The point is made in Chapter 2 that IQ heritability (h^2) as such (either "broad" or "narrow" heritability) has no implications for educational policy, although "environmentability" (e^2) does.
2. There is evidence that such an impact can go in either direction, one that might loosely be interpreted as "hereditarian" in policy orientation and the other "environmentalist." On the hereditarian side, it has been noted that the research of Cyril Burt, upon which Jensen originally based much of his own argument, was significantly responsible for the establishment of England's tracking system, which stratifies students into educational categories at an early age. According to Gillie (1977, p. 469, reprinted from the October 27, 1976, London *Sunday Times*), and Hearnshaw (1979), Burt strongly influenced Britain's 1944 Education Act, which established grammar, technical, and secondary schools. In the United States, Jensen's 1969 article has been cited in a number of local United States court cases as justification for educational segregation of races (C. Martin, personal communication, 1975). On a more environmentalist side, as a result of a well-publicized California case (*Larry P. et al.* v. *Wilson Riles et al.*), IQ tests were banned in 1971 in San Francisco as a method of classifying (black) students as "educable mentally retarded." In 1974 the ban became statewide (see Wade, 1978). Supreme Court cases (e.g., *Griggs et al.* v. *Duke Power Company*) have placed limits on the uses of a number of psychological tests, and vigorous policy-oriented criticisms (Rodriguez, 1974; Stone, 1975) as well as threatened lawsuits mount against Educational Testing Service (ETS), the organization that administers such well-known achievement and aptitude tests as the Scholastic Aptitude Test (SAT), the Graduate Record Examination (GRE), and the Law School Admission Test (LSAT). Recent among these is the "Nader Report" on ETS (A. Nairn and associates, "The Reign of ETS: The Corporation that Makes Up Minds; the Ralph Nader Report on the Educational Testing Service," 1980).

3. My evaluation of Herrnstein's (1971, 1973) work will center primarily on his conclusion that IQ heritability is "substantial" (80%), a conclusion he bases almost exclusively on Jensen's 1969 article, which I consider in detail. His tendency to misread the literature he searched "from one end to the other" is exemplified by the following kinds of errors in his book (Herrnstein, 1973): In one place (p. 79), he implies that a high correlation means "direct causality"; in another (p. 177) he confuses main effects with interactions in analysis of variance; and in another (p. 82) he implies that a correlation of 1.00 between X and Y means that a numerical score on X necessarily translates directly into the same numerical score on Y. In other places, he makes the all-too-common errors of equating "education" with "social environment" (p. 148 and passim), "occupation" with "socio-economic status" (p. 160 and passim), and "socio-economic status" with "social environment" (pp. 159–60 and passim).

4. Like Herrnstein (1973; see note 3), Eysenck misreads and misinterprets the literature—sometimes even misreading his own tables. For example, on p. 103 (Eysenck, 1971) he reverses the interpretation of a relationship, stating that black-white differences in IQ increase as SES increases, while he should have stated that the differences *decrease* (as shown in a table in Jensen, 1969, p. 83, to which Eysenck is presumably referring). On p. 53, he misinterprets his own figure (Figure 7b, source uncited) on the relationship between IQ and EEG wave length, stating that the relationship is inverse whereas *within* both the "high IQ" and the "low IQ" categories the relationship appears direct. On p. 61, he subsumes gene-environment interaction under "genetic" variance, a bizarre procedure followed by virtually no other scholar, which violates all logic of multivariate analysis. On p. 91 and passim he confuses "nonwhite" with "black." In numerous places he cites a table without citing either a secondary or a primary source (e.g., p. 22, p. 53, p. 58, p. 81, p. 118, p. 122, to name only some).

CHAPTER 2

1. Certain errors appear in Jensen's statement of equations *(2.1)* and *(2.6)* in his article in the *Harvard Educational Review* (1969, p. 34), where an equation is published as

$$V_p = \frac{(V_G + V_{AM}) + V_D + V_i}{V_H} + \frac{V_E + 2\text{Cov}_{HE} + V_I}{V_E} + V_e,$$

which is incorrect, since one does not divide by V_H or by V_E. The division lines were meant as brackets, such that $V_H = V_G + V_{AM} + V_D + V_i$ and $V_E = V_E + 2\text{Cov}_{HE} + V_I$. (This assumes, however, that both covariance and interaction are to be subsumed under environmental variance.) This error was corrected in later publications (e.g., Jensen, 1970, 1973). A second error appears in Jensen (1973, p. 367): The covariance term is given as "Cov_{GE}" rather than as "2Cov_{GE}," which is correct.

2. After publication of Jensen's 1969 article, some initial criticism implied that r_{MZA} should itself be squared (r^2_{MZA}) to estimate h^2, since most people in the behavioral sciences are taught to square a correlation coefficient in order to obtain "explained variance." This would of course give a lesser value for the estimate of h^2, since $r^2_{MZA} < r_{MZA}$ if $1 > r_{MZA} > 0$. But Jensen is correct here—if a flagrantly long list of arbitrary assumptions is made. Jensen (1971) gives and repeats (1976) a proof of sorts, but without listing most of the requisite assumptions. I derive $h^2 = r_{MZA}$ in Chapter 4 to illustrate the requisite assumptions in a more formal and complete way.

3. A path analysis using the case of siblings (or dizygotic twins) illustrates how a correlation for the genotypes of each pair member (r_{G1G2}) is derived. If the parents' genotypes (G_M, G_F) affect offspring genotypes (G_1 and G_2) for two of their own biological offspring (and no environmental variables affect G_1 and G_2), a simple path diagram, not considering phenotypes, can be created:

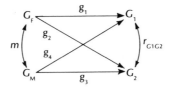

Path coefficients $g_1 = g_2 = g_3 = g_4 = \frac{1}{2}$. If $m = 0$ (if no correlation between parental genotypes arises because of assortative mating), it follows directly from the path theorem that

$$r_{G1G2} = (\tfrac{1}{2})^2 + (\tfrac{1}{2})^2 = \tfrac{1}{2} = .5,$$

the expected genotypic correlation for one pair of siblings (and DZs) under the simplest genetic model (Table 2.2, last column). Note that if assortative mating has resulted in some correlation of parental genotypes, $m > 0$ and thus $r_{G1G2} > .5$ (Table 2.2, third column). Path models are discussed in detail in Chapters 4 and 5.

4. Holzinger's (1929) study of 25 MZTs and 26 DZTs is not independent of the Newman et al. (1937) 50 MZTs and 50 DZTs; the former samples are part of the latter. Jencks et al. acknowledge this in a footnote (p. 317) but appear nevertheless to treat them as separate samples in their compilation. For the moment, they are so treated here. Later Jencks et al. do indeed use only Newman et al. (1937) for estimates of the IQ correlation for MZTs and DZTs.

5. This is discussed in more detail in Chapter 4. Briefly, given a large number of assumptions, one can derive (from path analysis) the following basic equation:

$$r_{UNT} = e^2 r_{E1E2} + h^2 r_{G1G2} + 2her_{GE},$$

where r_{E1E2} is the intrapair environmental similarity (environmental correlation); r_{G1G2} is genetic similarity (ρ in Jensen's equation); and r_{GE} is the gene-environment correlation. For unrelated persons, where both are adopted, one can assume that $r_{G1G2} = r_{GE} = 0$, and thus:

$$r_{UNT} = e^2 r_{E1E2}.$$

Nonfamily (residual) variance is represented by u^2, and

$$r_{UNT} + u^2 = 1.$$

Thus

$$1 = e^2 r_{E1E2} + u^2,$$

showing that r_{UNT} subsumes only family environmental variance. Note that $e^2 = r_{UNT}$ only if $r_{E1E2} = 1$, a doubtful assumption.

6. The predictions that

$$h^2(\text{MZT, SIBT}) > h^2(\text{MZT, DZT})$$

and that

$$h^2(\text{DZT, UNT}) > h^2(\text{SIBT, UNT}),$$

can be understood by comparing appropriate equations derivable, after many assumptions, from path analysis (as well as from variance procedures). These equations (which we review in Chapter 4) are:

$$r_{MZT} = h^2(1) \quad + e^2 r_{E1E2(MZT)} + 2her_{GE(MZT)} \tag{1}$$

$$r_{DZT} = h^2(.55) + e^2 r_{E1E2(DZT)} + 2her_{GE(DZT)} \tag{2}$$

$$r_{SIBT} = h^2(.55) + e^2 r_{E1E2(SIBT)} + 2her_{GE(SIBT)} \tag{3}$$

$$r_{UNT} = 0 \qquad + e^2 r_{E1E2(UNT)} + 0 \tag{4}$$

Here, the r_{E1E2} quantities represent the environmental correlation (intrapair environmental similarity) for each kinship category. The genetic correlation (intrapair genetic similarity) is 1.00 for MZs and .55 for DZs and SIBs. The r_{GE} quantities represent the gene-environment correlations for each kinship. I use UNT-AA (rather than UNT-AN or UNT-P) for illustration. If the gene-environment correlations are assumed to be nonzero but equal across kinship categories such that $r_{GE(MZT)} = r_{GE(DZT)} = r_{GE(SIBT)}$, then by subtracting selected pairs of equations, one can see what the comparisons of interest subsume. Subtracting equation (2) from equation (1) (for comparing MZT to DZT) and equation (3) from equation (1) (for comparing MZT to SIBT), produces

$$(r_{MZT} - r_{DZT}) = h^2(1 - .55) + e^2(r_{E1E2(MZT)} - r_{E1E2(DZT)}) + 0 \tag{5}$$

$$(r_{MZT} - r_{SIBT}) = h^2(1 - .55) + e^2(r_{E1E2(MZT)} - r_{E1E2(SIBT)}) + 0. \tag{6}$$

Subtracting equation *(5)* from equation *(6)*, h^2 drops out, leaving

$$[(r_{MZT} - r_{SIBT}) - (r_{MZT} - r_{DZT})]$$
$$= e^2[(r_{E1E2(MZT)} - r_{E1E2(SIBT)}) - (r_{E1E2(MZT)} - r_{E1E2(DZT)})].$$

This shows that if $(r_{MZT} - r_{SIBT})$ exceeds $(r_{MZT} - r_{DZT})$, which is in fact the case, then for given (positive) values of e^2, the only source of this difference is a positive difference between the excess environmental correlation of MZTs over SIBTs and the excess environmental correlation of MZTs over DZTs. This is empirical evidence that, as suspected (assuming equal gene-environment correlations), $r_{E1E2(DZT)} > r_{E1E2(SIBT)}$, that is, that intrapair environmental similarity for DZTs is greater than that for SIBTs, even though both of these environmental correlations are unmeasured, unobserved quantities. Similar reasoning applies to the comparisons of DZTs and UNTs to SIBTs and UNTs. Note, incidentally, that unlike the above, in the eight other such "difference of difference" comparisons h^2 (and r_{G1G2}) enter into the comparison. See Chapter 4 for more details.

7. Again, comparisons of pairs of kinship equations (see note 6) permit this to be seen more clearly.

8. Although the DeFries formula uses four quantities (h_B^2, h_W^2, t, and r), if race (R) and "intelligence genotype" (G) are thought of as continuous variables, then the Yule and Kendall (1950, p. 301) "consistency equation" (cf. Costner and Leik, 1964), which uses only three quantities, applies. Let h^2 represent the (squared) product-moment correlation between G and IQ for all persons of all races. Let t' represent the product-moment correlation between R and IQ, and let r' represent the product-moment correlation between R and G. Then without further assumptions, if

$$h^2 + t'^2 \geq 1,$$

then $r' > 0$.

9. Given the four variables G, G', E, and Y (representing measured IQ), and assuming that all four are intercorrelated, this may be represented by the following set diagram:

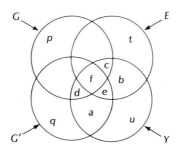

The variables (constructs) G, G', E, and Y are total variances. All intersections represent shared variances (covariances), and areas p, q, t, and u represent residual (unexplained) variances. Area a represents variance in Y that is explained by G' with both G and E constant. Area a would thus be given the intriguing interpretation that it represents variance in IQ that is explained by genotypes that are not "for" IQ. Area b represents variance in Y explained by E with both G and G' constant. Thus, u is that portion of the total IQ variance that is unexplained by G, G', E, and their covariances (such as variance explained by interactions). There are thus three distinct covariances that can explain portions of Y's total variance:

> areas $c + f$ (explained by covariance between G and E);
> areas $d + f$ (explained by covariance between G and G');
> areas $e + f$ (explained by covariance between G' and E).

A partitioning of all effects can be accomplished by means of path analysis (see Chapter 4). Assume the following path model:

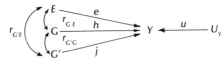

The coefficients h, e, and j are path coefficients (standardized partial regression coefficients), and coefficients r_{GE}, $r_{G'G}$, and $r_{G'E}$ are zero-order correlation coefficients. Note that squaring h (that is, h^2) results in broad heritability expressed as a path coefficient (the effect of G on Y with both E and G' constant). Coefficient j (or j^2) is thus given the interesting interpretation that it represents the effect upon IQ of genes that are not "for" IQ. Variable U_Y is Y's residual, and u^2 is thus that portion of Y's variance that is unexplained by G, G', and E. It can be shown that the following partitioning results from application of the path theorem:

$$1 = \frac{Var_Y}{Var_Y} = h^2 + e^2 + j^2 + 2her_{GE} + 2hjr_{GG'} + 2ejr_{G'E} + u^2,$$

where $2her_{GE}$ is the proportion of Y's variance that is explained by covariance between G and E (corresponding to areas $c + f$ in the set diagram above); $2hjr_{GG'}$ is that proportion explained by G and G' covariance (areas $d + f$); and $2ejr_{G'E}$ is that proportion explained by G' and E covariance (areas $e + f$). If all assumptions implicitly made here are correct (that G and G' are indeed correlated; that additivity exists; etc.), then there are seven distinct sources of variance for IQ, and thus a total of seven unknowns, namely: h^2, e^2, j^2, the three covariances, and u^2. All components necessarily sum to 1.00 given the inclusion of a residual for Y.

10. I am indebted to Professor Michael Schwartz for calling my attention to the possibility of G-E interaction having a dual character analogous to that of G, E covariance.

CHAPTER 3

1. Kamin's (1974) innovative analysis of these same data stresses the irregularities in Burt's data (see Chapter 2) as well as the issue of measurement error (especially the confounding of age with IQ score, an issue mentioned later in this chapter). In this chapter, I focus upon the effect of similarities in the separated twins' social environments on their IQs for the three studies for which data are available (those of Shields, Newman et al., and Juel-Nielsen). Kamin's treatment of environmental similarity, while remarkably insightful, is unsystematic and incomplete. My analysis differs from Kamin's in the following respects: (1) Kamin analyzes the effect of relatedness of adoptive families only for the Shields study, whereas I analyze this for the other two studies as well. (2) I analyze separately effects of four variables—age of the twins at separation, whether or not they were reunited after their initial separation, relatedness of adoptive families, and similarities in social and educational environment after their presumed separation. Kamin does not attempt to treat these environmental variables separately. (3) It is useful (though not always advisable) to compare averaged values of IQ correlations across studies. I calculate both separate and averaged values for a given environmental variable across studies; Kamin employs neither procedure. (4) Kamin bases his statistical tests of significance upon Ns of single MZ individuals rather than upon Ns of pairs, which is the correct procedure. (It is easier to establish significance with individuals, since individual Ns are twice as large as pair Ns. In general, the likelihood of establishing statistical significance increases as N increases.) I base significance tests upon pairs. (5) I employ a path analysis model from Jencks et al. (1972) to reevaluate Jencks et al.'s analysis of the Newman et al. separated twins. I obtain heritability estimates considerably below those of Jencks et al. (including some estimates of zero heritability).

2. A total N of 122 has been made quite famous by Jensen (1970b), because he included 38 pairs from Shields rather than the 37 actually analyzed by Shields. The extra pair are named Herbert and Nicholas, Shields's case SM4, unused by Shields since both twins had been in mental hospitals for long periods, and Shields regarded their test scores as unreliable. Jensen, however, (1970b) included their IQ scores in his analysis (32 and 26 raw; 75 and 71 transformed). Their transformed scores thus differ by 4 points, less than the mean of the intrapair differences for the entire sample (7.68). Inclusion of this case would thus slightly increase the overall IQ correlation, r_{MZA} (from .77 to .78). I use only the 37 pairs analyzed by Shields himself.

3. A handful of other case studies of presumably monozygotic twins raised separately is scattered in some of the earlier literature. Evidently, the earliest published account of a separated pair tested for IQ was that of Muller (1925). A British pair reared apart up to the age of 10 years is described in

Saudek (1934). After publication of their work in 1937, Newman et al. added a twentieth pair to their sample, described in Gardner and Newman (1940). A brief summary of these single-pair studies appears in Shields (1962, pp. 9–16). Kamin (1978) has noted that four pairs were reported in detail in Burks and Roe (1949).

4. There is actually a fourth condition: that any prenatal environmental (intrauterine) effects upon the twins' IQs are equal and also negligible in magnitude. More will be said about intrauterine effects later.

5. Of the three, Shields (1962, pp. 62, 101, and passim) makes the best attempt to determine how IQ similarity (as measured by mean intrapair IQ differences) is affected by such variables as whether or not the twins attended the same school. He shows that these variables do have some effect on the mean intrapair IQ differences. Shields makes no significance tests however, nor does he calculate separate IQ correlations for the different categories of the environmental variables.

6. In my analysis both the composite raw and transformed scores from Shields are used. Shields's raw scores were transformed as follows: Given Shields's estimates, an IQ of 100 corresponds to a raw Dominoes score of 28 points and a raw Mill Hill score of 19 points, thus giving a composite raw score of 28 + 19(2) or 66 points. The sample standard deviation is 18.36, across his 37(2) or 74 separate scores. The transformed "IQ" score for a single individual to achieve a mean of 100 and a standard deviation of 15 is thus

$$\text{IQ score}_i = \frac{(x_i - 66)15}{18.36} + 100,$$

where x_i = the raw composite score for the ith individual. Transforming the *composite* scores seemed preferable to making separate transformations for the Dominoes and the Mill Hill (as done by Jensen, 1970b), given Shields's own scoring procedure (first doubling the Mill Hill score), and treating the raw scores as given. One transformation should introduce less distortion than two transformations. Nonetheless, both raw and transformed scores are separately analyzed later. Differences are small, but a slight tendency for transformed scores to give a higher (inflated) IQ correlation (r_{MZA}) exists.

7. The expression 1.13σ comes from Gini's formula, from Kendall (1960), cited in Jensen (1967, p. 301), where $|\bar{d}_p|$, or $|\bar{x}|$, is $2\sigma/\sqrt{\pi}$; $\sqrt{\pi} = 1.772$, thus $|\bar{x}| = 2\sigma/1.772 = \sigma(2/1.772) = \sigma 1.13$. Jencks et al. also use $\sigma 1.13$ as an estimate of the mean absolute difference for all possible paired comparisons in a population (1972, p. 311).

8. As Kamin (1974, p. 178) notes, there is literally no psychology student in America who has not been exposed to at least part of John B. Watson's behaviorism: "Give me a dozen healthy infants, well-formed, and my own specified world to bring them up in and I'll guarantee to take anyone at random and train him to become any type of specialist I might select—

doctor, lawyer, artist, merchant-chief and yes, even beggar-man and thief, regardless of his talents, penchants, tendencies, abilities, vocations, and race of his ancestors" (Watson, 1930, p. 103).

9. Correlations, mean absolute intrapair differences, means, and standard deviations were calculated for all MZA pairs for each study, with the following results. These statistics all agree exactly with the original sources:

	r	N	Absolute difference	Mean	s.d.
Shields					
Raw scores	.77	37	9.46	56.84	18.36
Transformed scores	.77	37	7.68	92.57	14.97
Newman et al.	.67	19	8.21	95.68	13.00
Juel-Nielsen					
Time$_1$ scores	.62	12	7.25	105.54	9.42
Combined scores	.69	12	6.42	106.79	9.09
Averaged correlation	.73	68			

10. See note 3.
11. Twins who were never really separated in the first place (such as twins living next door to one another) were coded as "reunited" for purposes of analysis. "Reunion" was coded separately from "similarity in social environment," analyzed later. (As the table in note 12 shows, several twins across all three studies were classified as reunited but as not having "similar" social environments.)
12. MZ pairs classified as "minimal" in similarity in social environment can at the same time have other environmental characteristics in common (childhood reunion; related adoptive families), as the following table (adding cases across studies) makes clear:

Childhood reunion	Yes		No		
Related families	Definite	Possible	Definite	Possible	Total
Similarity in social environment					
Strong	17	2	1	0	20
Possible	16	1	6	3	26
Minimal	5	3	3	11	22
Total	38	6	10	14	68

The marginal totals correspond to appropriate totals in Tables 3.1 through

3.4. The three possible bivariate classifications show very marked, though not perfect, relationships among the three environmental variables (note that the unit is the *pair*): reunion by family relatedness ($\chi^2 = 15.23$, $P <$ .001); reunion by similarity in social environment ($\chi^2 = 18.66$, $P <$ .001); and family relatedness by similarity in social environment ($\chi^2 = 18.35$, $P <$ .001). Clearly, one cannot justifiably calculate IQ correlations (or perform analyses of variance [ANOVA] on intrapair IQ differences) on a multivariate classification, since over half of all subclasses in the above table contain too few cases (≤ 3 pairs). Note finally that the univariate consideration of each environmental variable (in Tables 3.1 through 3.4) is in effect an investigation of the assumption that the (bivariate) environmental correlation (the quantity r_{E1E2} in equation [3.2]) is zero or approximately so for "separated" MZ twins for any given environmental variable. (The quantity r_{E1E2} is formally introduced in Chapter 4.)

13. A. Mazur (personal communication, 1978) has raised the intriguing issue of whether the classification of pairs into "minimal" similarity in social environment might have resulted in a slightly negative correlation of environments. If this were the case, the environmental correlation quantity r_{E1E2} in equation *(3.2)* would be negative, thus overestimating h^2, holding all other terms, including $r_{Y1Y2(MZA)}$, constant. It would thus be on the conservative side (for an environmentalist) to estimate h^2 from equation *(3.2)*. (I would underestimate h^2 if only the IQ correlation itself were used as a direct estimate of h^2.) It was not possible to code E_1 reliably (e.g., occupation or education of the environment of twin$_1$) separately from E_2 (the same variables for twin$_2$); see note 12 and Appendix B. Judging from the descriptive data, such negative environmental correlation seems a remote possibility, but the issue is worth investigating further. This issue has also been raised by Jencks (personal communication, 1979).

14. Fulker's (1975, p. 519) value of .95 for the Newman et al. separated MZ twins is so far out of line with all other research that Goldberger was motivated (as am I) to suspect it is a "wholly invented figure" (1978b, p. 6). I wrote Professor Fulker for an explanation (not given in the article) of why this value was so absurdly high. His response (in a personal communication, 1979) indicated that he somehow "corrected" the twins' IQ differences "for differences in education and social background, using multiple regression." Why such a "corrected" value should be used as a datum is unclear. Furthermore, he admits himself that the .95 value "is rather a suspect statistic."

15. Derivation of path estimation equations directly from the path theorem is taken up in Chapter 4.

CHAPTER 4

1. The assumptions listed in the text are relatively standard in parametric multivariate analysis generally, and multiple regression and path-analytic

techniques, specifically. (While assumptions 6 and 7 might be construed as merely desirable rather than necessary, the rest are necessary in order to estimate h^2 for IQ.) For good general discussions, the reader is referred to Blalock (1972), Bohrnstedt and Carter (1971), and Kerlinger (1973). For discussions specific to the requirements and limitations of path analysis in the social and behavioral sciences, see Goldberger (1972), Goldberger and Duncan (1973), Hauser and Goldberger (1971), Heise (1969a), Kerlinger and Pedhazur (1973), Land (1969), Namboodiri et al. (1975), and Werts and Linn (1970). The reader is also referred to the seminal work of Blalock (for example, 1962, 1964, 1968, 1971), Boudon (1965), and Duncan (1966, 1969, 1970, 1975). For discussions of the kinds of assumptions specific to the analysis of the heritability of continuous human phenotypes (such as IQ scores), see Layzer (1974), Lewontin (1974, 1975), Morton (1972, 1974), De-Fries (1972), McClearn and DeFries (1973), Plomin et al. (1977), Burt and Howard (1956), and Newman et al. (1937, pp. 3–31). Throughout this chapter and the next, I also rely on the excellent critical discussions of Goldberger (especially 1976a, 1976b, 1976c, 1977, 1978a, 1978b, 1978c, 1978d).

2. Some notable exceptions are: Cavalli-Sforza and Feldman (1973); Jinks and Fulker (1970); Light and Smith (1969); and Rao and Morton (1974).

3. While reciprocal causation is routinely handled in some fields (such as econometrics), other fields (sociology, psychology, and evidently some of behavior genetics) seem to proceed as follows: Given three variables G (IQ genotype), E (environment), and Y (measured IQ score), and allowing *each* variable its own residual (U), a nonrecursive system of linear regression equations (permitting two-way causation for all pairs of variables) is

$$Y = \alpha_Y + \beta_{YY}Y + \beta_{YE}E + \beta_{YG}G + U_Y$$
$$E = \alpha_E + \beta_{EY}Y + \beta_{EE}E + \beta_{EG}G + U_E$$
$$G = \alpha_G + \beta_{GY}Y + \beta_{GE}E + \beta_{GG}G + U_G.$$

The β coefficients are regression coefficients, and the α are the Y-intercepts. This system is unmanageable (there are far too many unknowns), so arbitrary assumptions are made to eliminate certain of these unknowns, with the hope that no severe distortion of reality results. Following the usual conventions, Y, E, and G are assumed in standard form; thus $\alpha_Y = \alpha_E = \alpha_G = 0$, and the coefficients ($\beta$'s) become standardized partial regression coefficients (path coefficients). It is further assumed that U_Y, U_E, and U_G are uncorrelated (orthogonal) to each other and to the other two variables not subscripted by each. It is assumed still further that no variable "causes itself"; thus, the diagonal coefficients (β_{YY}, β_{EE}, and β_{GG}) are all zero and that recursiveness exists such that if $\beta_{ij} \neq 0$, then $\beta_{ji} = 0$; setting $\beta_{EY} = \beta_{GY} = \beta_{GE} = 0$ permits β_{YE}, β_{YG}, and β_{EG} to be nonzero. $\beta_{YG}^2 = h^2$ and $\beta_{YE}^2 = e^2$. The resulting system of equations, obtained after all these quite arbitrary assumptions, is recursive, contains far fewer unknowns, and is thus in principle closer to empirical solution:

$$Y = \beta_{YE}E + \beta_{YG}G + U_Y$$
$$E = \beta_{EG}G + U_E$$
$$G = U_G.$$

These equations correspond to the following causal model; compare this model to the models in Figure 4.1.

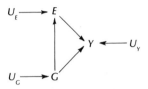

4. Certain models involving parent-child correlations, for example those of Rao et al. (1976), Rao and Morton (1978), and Jencks et al. (1972), provide for a causal path from the parent's IQ score to the child's environment. I take these models up in Chapter 5. But I know of no analysis in which the IQ of an individual affects his or her own environment (or that of the paired partner), although some (e.g., Eaves, 1976) have raised the issue conceptually.

5. The interested reader might consult note 3 in Chapter 2, as well as the parent-offspring equations reviewed in Chapter 5, to see how the genetic correlation r_{G1G2} can be derived for different kinships.

6. It seems to me that one does not get around the problem by defining some construct (r_{E1E2}) as representing only common environment, thus putting all noncommon (specific) environmental effects in the residual (along with everything else that is there) as do Jencks et al. (1972), Rao et al. (1976, 1978), Rao and Morton (1978), and Loehlin (1978). Doing so eliminates a priori the possibility of allowing for differences across kinships in r_{E1E2} as well as in the correlation r_{UY1UY2} (which is always assumed to be zero). Of what does common environment consist? Defining it as anything "that is common" is tautological. I retain the distinction between r_{E1E2} and r_{UY1UY2} throughout this book.

7. The quantity e^2r_{E1E2} is the proportion of IQ variance explained by intra-pair (within-family) environmental similarity (that is, between-family variance). Thus, under assumptions of additivity and linearity, u^2 is the proportion of IQ variance explained by nonfamily (between-family) environmental similarity (which is within-family variance). Assume, as does Jensen in his seminal article (1967) although the assumption is implicit and implausible, that $r_{E1E2} = 1$ for pairs raised together, and further, that $r_{GE} = 0$ (thus $2her_{GE} = 0$). If $r_{G1G2} = 1$ (as for MZ twins, or for a general population of unpaired individuals), then from equation (4.15),

$$1 = e^2 + h^2 + u^2,$$

corresponding to Jensen's partitioning of IQ variance into "within-family environment," "between family environment," and heritability. In the case of MZ twins, Jensen assumes u^2 to be negligible, which is implausible, since generally $u^2 = (1 - r_{MZT})$ and generally this $u^2 > 0$. The reader is again reminded of the numerous assumptions, unstated in Jensen (1967), that must be made explicit to arrive at this partitioning.

8. Equation *(4.16)* using r_{tt} in place of r_{Y1Y2} for a general population of unpaired individuals is used by Jensen (1976) and appears in Chapter 5.

9. If adopted pairs (such as truly separated MZAs) indeed have zero gene-environment covariance such that $r_{GE(MZA)} = 0$ (just like adopted-adopted pairs, i.e., UNT-AA), then assuming other components constant, the total variance for MZAs (and any adopted pairs) will be reduced, that is, restricted. Rao et al. (1976) introduce a correction for this (their θ) which I consider in Chapter 5.

10. Subtracting equation *(4.30)* (for SIBTs) from *(4.29)* (for DZTs) gives

$$r_{Y1Y2(DZT)} - r_{Y1Y2(SIBT)} = e^2(r_{E1E2(DZT)} - r_{E1E2(SIBT)})$$

that is,

$$\Delta r_{Y1Y2} = e^2 \Delta r_{E1E2},$$

and thus

$$e^2 = (\Delta r_{Y1Y2})/(\Delta r_{E1E2}),$$

thus showing that for given values of Δr_{Y1Y2}, then Δr_{E1E2} determines e^2. Taking $\Delta r_{Y1Y2} = .10$, if (for example) $\Delta r_{E1E2} = .10$, then $e^2 = 1.00$; if $\Delta r_{E1E2} = .20$, then $e^2 = .50$; if $\Delta r_{E1E2} = .60$, which is implausibly high, then $e^2 = .17$. If lower values for Δr_{E1E2} are regarded as more plausible (as in much of the hereditarian literature), then a high e^2 is implied by even small differences (Δr_{Y1Y2}) in the IQ correlations of DZ twins and siblings. Burt, Erlenmeyer-Kimling and Jarvik, Jensen, and Jencks et al. do not appear to have considered the implications of this, nor does the Honolulu group, which excludes DZTs altogether from its analyses in Rao et al. (1976) and Rao and Morton (1978).

11. Equation *(4.37)* was obtained by Goldberger (1977), who explores the sensitivity of h^2 to differences in Δr_{E1E2} for MZT-DZT comparisons alone.

12. It is tempting to let f here stand for "finagle factor."

13. The effects of assortative mating on r_{G1G2} can be clearly seen from path models for direct line (parent-offspring) kinships, as in Figure 5.2 with variables (G, E, and Y) for two offspring included. See also note 3 in Chapter 2.

CHAPTER 5

1. Writing equations *(5.6)* through *(5.8)* to include the specified parameter values yields

$$213.75 = \sigma_E^2(1) + \sigma_G^2(1) + 2\sigma_G\sigma_E r_{GE} \text{ (General population)} \quad (5.6')$$

$$195.75 = \sigma_E^2(.9) + \sigma_G^2(1) + 2\sigma_G\sigma_E r_{GE} \quad \text{(MZTs)} \quad (5.7')$$

$$126.00 = \sigma_E^2(.6) + \sigma_G^2(.5) + 2\sigma_G\sigma_E r_{GE}. \quad \text{(DZTs)} \quad (5.8')$$

Subtracting equation (5.8') from (5.6') to eliminate $2\sigma_G\sigma_E r_{GE}$, and (5.7') from (5.6') to eliminate σ_G^2, gives the following equivalent system

$$213.75 = \sigma_E^2(1) \qquad + \sigma_G^2(1) \qquad + 2\sigma_G\sigma_E r_{GE} \quad (1)$$

$$87.75 = \sigma_E^2(1-.6) + \sigma_G^2(1-.5) + 0 \quad (2)$$

$$18.00 = \sigma_E^2(1-.9) + 0 \qquad + 0. \quad (3)$$

From equation (3), $18.00 = \sigma_E^2(.1)$; thus $\sigma_E^2 = 180.00$. Substituting this value for σ_E^2 into equation (2) to solve for σ_G^2 yields $87.75 = 72 + \sigma_G^2(.5)$; thus $\sigma_G^2 = 31.50$. Both σ_E^2 and σ_G^2 are substituted into equation (1) to obtain: $2.25 = 2(31.50)^{\frac{1}{2}}(180.00)^{\frac{1}{2}}(r_{GE})$; thus $r_{GE} = .0149$. All these values satisfy original equations (5.6')–(5.8'), and the solution is unique. (One of course gets the same results if one writes equations [5.6']–[5.8'] in terms of the raw correlations [.95, .87, .56] and quantities h^2, e^2, and r_{GE}.)

2. The misspecification involved using, for kin (and equation) 8 in Figure 5.1, the expression $i(p + qa)$, which refers to the correlation of parent's IQ with an environmental index of his or her own childhood environment, rather than $i(t)$, that is,

$$i[f(p(1 + u_6) + qa) + x(1 + p^2 u_6)],$$

which refers, correctly, to a parent's IQ with the child's environmental index (Goldberger, 1978d; 1978b, pp. 56–61; 1978c, pp. 201–3).

3. The program used, however, minimized the difference between the z-transforms of observed and expected correlations:

$$z_i = (\tfrac{1}{2})\log[(1 + r_i)/(1 - r_i)],$$

where the r_i are correlations, and where the weights are the n_i. I thank Robert Porter of Princeton University's department of economics for his able assistance in adapting the GQ OPT program and accomplishing all data runs reported here.

4. Extending our notational scheme, the Birmingham specification for r_{G1G2} for siblings and DZs (after Goldberger, 1978b, p. 18) is as follows:

$$r_{G1G2} = \tfrac{1}{4} + \tfrac{1}{2}D/(1 + F) + 2h^2 E^2 F/(1 + F)^2,$$

where $D = r_{G1G2(P,O)}$ = the genetic correlation for parent and offspring;
$F = r_{Y1Y2(M,F)}$ = the phenotypic correlation between spouses (Honolulu's r_{FMT}); and
$E = r_{E1E2(P,O)}$ = the environmental correlation between parent and offspring (Honolulu's $r_{OPTIXIY}$).

CHAPTER 6

1. This point is noted, in almost the same words, by philosophers Block and
   Dworkin (1976, p. 479) in their excellent discussion of intelligence, IQ
   measurement, and the meaning of heritability. The point seems shared in
   spirit as well in the treatments of others such as Segall (1976) and Mazur
   and Robertson (1972, pp. 86–89).

References

Althauser, R. P. 1971. Multicollinearity and non-additive regression models. In *Causal models in the social sciences,* ed. H. M. Blalock, Jr., pp. 453–72. Chicago: Aldine.

Babson, S. G.; Kangas, J.; Young, N.; and Bramhill, J. L. 1964. Development of twins of dissimilar size at birth. *Pediatrics* 33: 329–30.

Baldwin, A. L. 1969. A cognitive theory of socialization. In *Handbook of socialization theory and research,* ed. D. A. Goslin, pp. 325–45. Chicago: Rand-McNally.

Bandura, A. 1969. Social learning theory of identificatory processes. In *Handbook of socialization theory and research,* ed. D. A. Goslin, pp. 213–62. Chicago: Rand-McNally.

Behrman, J., and Taubman, P. 1976. Intergenerational transmission of income and wealth. *American Economic Review* 66: 436–40.

Behrman, J.; Taubman, P.; and Wales, T. 1977. Controlling for and measuring the effects of genetic and family environment in equations for schooling and labor market success. In *Kinometrics: Determinants of socioeconomic success within and between families,* ed. P. Taubman, pp. 35–96. Amsterdam: North-Holland.

Benton, A. L., and Van Allen, M. W. 1971. Prosopagnosia and facial discrimination. *Journal of the Neurological Sciences* Elsevier Publishing Co., Amsterdam (June).

Blalock, H. M., Jr. 1962. Four-variable causal models and partial correlations. *American Journal of Sociology* 68: 182–94.

———. 1964. *Causal inferences in nonexperimental research.* Chapel Hill: University of North Carolina Press.

———. 1965. Theory building and the statistical concept of interaction. *American Sociological Review* 30: 374–80.

———. 1966. The identification problem and theory building: The case of status inconsistency. *American Sociological Review* 31: 52–61.

———. 1967a. Status inconsistency and interaction: Some alternative models. *American Journal of Sociology* 73: 305–15.

244

————. 1967b. Status inconsistency, social mobility, status integration, and structural effects. *American Sociological Review* 32: 780–801.

————. 1968. Theory building and causal inferences. In *Methodology in social research,* ed. H. Blalock, Jr. and A. B. Blalock, pp. 155–98. New York: McGraw-Hill.

————. 1969. *Theory construction: From verbal to mathematical formulation.* Englewood Cliffs, N.J.: Prentice-Hall.

————. 1970. A causal approach to nonrandom measurement errors. *American Political Science Review* 64: 1099–1111.

————. ed. 1971. Causal models in the social sciences. Chicago: Aldine.

————. 1972. *Social statistics.* 2d ed. New York: McGraw-Hill.

Blewett, D. B. 1954. An experimental study of the inheritance of intelligence. *Journal of Mental Science* 100: 922–33.

Block, N. J., and Dworkin, G., eds. 1976. *The IQ controversy: Critical readings.* New York: Pantheon Books.

Bogen, J. E. 1969a. The other side of the brain I: Dysgraphia and discopia following cerebral commissurotomy. *Bulletin of the Los Angeles Neurological Society* 34: 73–107.

————. 1969b. The other side of the brain II: An appositional mind. *Bulletin of the Los Angeles Neurological Society* 34: 135–62.

Bogen, J. E., and Bogen, G. M. 1969. The other side of the brain III: The corpus callosum and creativity. *Bulletin of the Los Angeles Neurological Society* 34: 191–220.

Bogen, J. E., and Gazzaniga, M. S. 1965. Cerebral commissurotomy in man: Minor hemisphere dominance for certain visuospatial functions. *Journal of Neurosurgery* 23: 394–99.

Bogen, J. E.; DeZure, R.; TenHouten, W. D.; and Marsh, J. F. 1972. The other side of the brain IV: The A/P ratio. *Bulletin of the Los Angeles Neurological Society* 37: 49–61.

Bohrnstedt, G. W. 1976. The reliability of product variables. Paper read at the Socio-Environmental Studies Laboratory, National Institutes of Health, Washington, D.C.

Bohrnstedt, G. W., and Carter, T. M. 1971. Robustness in regression analysis. In *Sociological methodology 1971,* ed. H. L. Costner, pp. 118–46. San Francisco: Jossey-Bass.

Bohrnstedt, G. W., and Marwell, G. 1978. The reliabilities of products of two random variables. In *Sociological methodology 1978,* ed. K. Schuessler, pp. 254–73. San Francisco: Jossey-Bass.

Boring, E. G. 1923. Intelligence as the tests test it. *New Republic* 34: 35–36.

Boudon, R. 1965. A method of linear causal analysis: Dependence analysis. *American Sociological Review* 30: 365–74.

Bradshaw, J. L. 1977a. Role of the cerebral hemispheres in music. *Brain and Language* 4: 403–31.

————. 1977b. Music perception and cerebral asymmetries. *Cortex* 13: 390–401.

———. 1978. Visual field differences in verbal tasks: Effects of task familiarity and sex of subject. *Brain and Language* 5: 166–87.

Bradshaw, J. L., and Gates, E. A. 1975. A note on interactions of ears, hands and cerebral hemispheres in a musical performance task requiring bimanual coordination. *British Journal of Psychology* 66: 165–68.

Bradshaw, J. L.; Gates, E. A.; and Nettleton, N. C. 1971. Bihemispheric involvement in lexical decisions: handedness and a possible sex difference. *Neuropsychologia* 15: 277–86.

Bradshaw, J. L.; Gates, E. A.; and Patterson, K. 1976. Hemispheric differences in processing visual patterns. *Quarterly Journal of Experimental Psychology* 28: 667–81.

Bronfenbrenner, U. 1975. Nature with nurture: A reinterpretation of the evidence. In *Race and IQ*, ed. A. Montague, pp. 114–44. New York: Oxford University Press.

Burks, B. S. 1928. The relative influence of nature and nurture upon mental development: A comparative study of foster parent–foster child resemblance and true parent–true child resemblance. In *Twenty-seventh yearbook of the national society for the study of education*, pt. 1, pp. 219–316. Bloomington: Public School Publishing Co.

Burks, B. S., and Roe, A. 1949. Studies of identical twins reared apart. *Psychological Monographs* 63(300): 1–62.

Burt, C. 1921. *Mental and scholastic tests*. London: King.

———. 1933. *Handbook for tests for use in schools*. London: Staples.

———. 1955. The evidence for the concept of intelligence. *British Journal of Educational Psychology* 25: 158–77.

———. 1958. The inheritance of mental ability. *American Psychologist* 13: 1–15.

———. 1961. Intelligence and social mobility. *British Journal of Statistical Psychology* 14: 3–24.

———. 1966. The genetic determination of differences in intelligence: A study of monozygotic twins reared together and apart. *British Journal of Psychology* 57: 137–53.

———. 1972. Inheritance of general intelligence. *American Psychologist* 27: 175–90.

Burt, C., and Howard, M. 1956. The multifactorial theory of inheritance and its application to intelligence. *British Journal of Statistical Psychology* 9: 95–131.

———. 1957a. Heredity and intelligence: A reply to criticisms. *British Journal of Statistical Psychology* 10: 33–63.

———. 1957b. The relative influence of heredity and environment on assessments of intelligence. *British Journal of Statistical Psychology* 10: 99–104.

Carmon, A., and Nachson, J. 1971. Effect of unilateral brain damage on perception of temporal order. *Cortex* 7: 410–18.

Cavalli-Sforza, L. L., and Feldman, M. W. 1973. Cultural versus biological inheritance: Phenotypic transmission from parents to children (a theory of the effect of parental phenotypes on children's phenotypes). *American Journal of Human Genetics* 25: 618–37.

Cohen, D. J.; Dibble, E.; Grawe, J. M.; and Polling, W. 1973. Separating identical from fraternal twins. *Archives of General Psychiatry* 29: 465–69.

Cohen, J. 1968. Multiple regression as a general data-analytic system. *Psychological Bulletin* 70: 426–43.

Cohen, R. A. 1967. *Primary group structure, conceptual styles, and school achievement.* Monograph, Learning Research and Development Center, Univ. of Pittsburgh, studies 1–4.

————. 1968. The relation between socio-conceptual styles and orientation to school requirements. *Sociology of Education* 41: 201–20.

————. 1969. Conceptual styles, culture conflict, and non-verbal tests of intelligence. *American Anthropologist* 51: 828–56.

Cohen, R. A.; Fraenkel, G.; and Brewer, J. 1968. The language of the hard core poor: Implications for culture conflict. *Sociology Quarterly* 10: 19–28.

Cole, S. 1979. *The sociological orientation.* Chicago: Rand McNally.

Conrad, H. S., and Jones, H. E. 1940. A study of familial resemblance in intelligence: Environmental and genetic implications of parent-child and sibling correlations in the total samples. In *Thirty-ninth yearbook of the national society for the study of education,* pt. 2, pp. 97–141. Bloomington: Public School Publishing Co.

Conway, J. 1958. The inheritance of intelligence and its social implications. *British Journal of Statistical Psychology* 11: 171–90.

————. 1959. Class differences in general intelligence: II. *British Journal of Statistical Psychology* 12: 5–14.

Costner, H. L. 1969. Theory, deduction, and rules of correspondence. *American Journal of Sociology* 75: 245–63.

Costner, H. L., and Leik, R. K. 1964. Deductions from axiomatic theory. *American Sociological Review* 29: 819–35.

CRM. *Society today.* 1973. Del Mar, Cal.: CRM Books.

Cromie, W. J. 1978. Male and female brains are different. In World Section of the *San Francisco Sunday Examiner and Chronicle* (January 22).

Cronin, J.; Daniels, N.; Hurley, A.; Kroch, A.; and Webber, R. 1975. Race, class, and intelligence: A critical look at the IQ controversy. *International Journal of Mental Health* 3: 46–132.

Dasen, P. R. 1972. The development of conservation in aboriginal children: A replication study. *International Journal of Psychology* 7: 75–85.

DeFries, J. C. 1972. Quantitative aspects of genetics and environment in the determination of behavior. In *Genes, environment, and behavior: Implications for educational policy,* ed. L. Ehrman, G. S. Omenn, and E. Caspari, pp. 5–16. New York: Academic Press.

DeLemos, M. M. 1969. The development of conservation in aboriginal children. *International Journal of Psychology* 4: 255–69.

DeRenzi, E., and Spinnler, H. 1966. Visual recognition in patients with unilateral cerebral disease. *Journal of Nervous Mental Disorders* 142: 515–25.

Dillard, J. L. 1972. *Black English: Its history and usage in the United States.* New York: Vintage Books-Random House.

Dobzhansky, T. 1973. *Genetic diversity and human equality: The facts and fallacies in the explosive genetics and education controversy.* New York: Basic Books.

Dobzhansky, T., and Spassky, B. 1944. Genetics of natural populations, XI. Manifestation of genetic variants in *Drosophila Pseudoobscura* in different environments. *Genetics* 29: 270–90.

Dorfman, D. D. 1978. The Cyril Burt question: New findings. *Science* 201: 1177–86.

———. 1979. Letter: Burt's tables. *Science* 204: 246–54.

Duncan, O. D. 1966. Path analysis: Sociological examples. *American Journal of Sociology* 72: 1–16.

———. 1969. Contingencies in constructing causal models. In *Sociological methodology 1969,* ed. E. F. Borgatta, pp. 74–118. San Francisco: Jossey-Bass.

———. 1970. Partials, partitions, and paths. In *Sociological methodology 1970,* ed. E. F. Borgatta and G. W. Bohrnstedt, pp. 38–47. San Francisco: Jossey-Bass.

———. 1975. *Introduction to structural equation models.* New York: Academic Press.

Duncan, O. D., and Featherman, D. L. 1973. Psychological and cultural factors in the process of educational achievement. In *Structural equation models in the social sciences,* ed. A. S. Goldberger and O. D. Duncan, pp. 229–53. New York: Seminar Press.

Duncan, O. D.; Featherman, D. L.; and Duncan, B. 1972. *Socioeconomic background and achievement.* New York: Academic Press.

Eaves, L. J. 1975. Testing models for variation in intelligence. *Heredity* 34: 132–36.

———. 1976. The effect of cultural transmission on continuous variation. *Heredity* 37: 41–57.

Eckland, B. K. 1967. Genetics and sociology: A reconsideration. *American Sociological Review* 32: 173–94.

———. 1971. Social class structure and the genetic basis of intelligence. In *Intelligence: Genetic and environmental influences,* ed. R. Cancro, pp. 65–76. New York: Grune and Stratton.

———. 1973. The biological basis of society. In CRM, *Society today.* Del Mar, Cal.: CRM Books.

Erlenmeyer-Kimling, L. 1972. Gene-environment interactions and the variability of behavior. In *Genes, environment, and behavior: Implications for educational policy,* ed. L. Ehrman, G. S. Omenn, and E. Caspari, pp. 181–209. New York: Academic Press.

Erlenmeyer-Kimling, L., and Jarvik, L. F. 1963. Genetics and intelligence: A review. *Science* 142: 1477–79.

Eysenck, H. J. 1967. *The biological basis of personality.* Springfield, Ill.: Charles C. Thomas.

————. 1971. *The IQ argument.* New York: Library Press.

————. 1977. The case of Sir Cyril Burt: On fraud and prejudice in a scientific controversy. *Encounter* 48: 19–24.

Eysenck, H. J., and Prell, D. B. 1951. The inheritance of neuroticism: An experimental study. *Journal of Mental Science* 97: 441–65.

Falconer, D. S. 1960. *Introduction to quantitative genetics.* New York: Ronald Press.

Feldman, M. W., and Lewontin, R. C. 1975. The heritability hang-up. *Science* 190: 1163–68.

Fernandez, R. 1977. *The I, the me, and you: An introduction to social psychology.* New York: Praeger.

Freeman, F. N.; Holzinger, K. J.; and Mitchell, B. C. 1928. The influence of environment on the intelligence, school achievement, and conduct of foster children. In *Twenty-seventh yearbook of the national society for the study of education*, pt. 1, pp. 103–217. Bloomington: Public School Publishing Co.

Fulker, D. W. 1975. Review of L. Kamin, *The science and politics of IQ. American Journal of Psychology* 88: 505–19.

Fuller, J. L. 1972. Discussion. In *Genetics, environment, and behavior: Implications for educational policy*, ed. L. Ehrman, G. S. Omenn, and E. Caspari, pp. 17–22. New York: Academic Press.

Galton, F. 1883. *Inquiries into the human faculty and its development.* London: Macmillan.

Gardner, I. C., and Newman, H. H. 1940. Mental and physical traits of identical twins reared apart: The twins Lois and Louise. *Journal of Heredity* 31: 119–26.

Gazzaniga, M. S. 1970. *The bisected brain.* New York: Appleton-Century-Crofts.

Gazzaniga, M. S.; Bogen, J. E.; and Sperry, R. W. 1962. Some functional effects of sectioning the cerebral commissures in man. *Proceedings of the National Academy of Sciences* 48: 1965–69.

Geffen, G.; Bradshaw, J. L.; and Nettleton, N. C. 1973. Attention and hemispheric differences in reaction time during simultaneous audio-visual tasks. *Quarterly Journal of Experimental Psychology* 25: 404–12.

Gillie, O. 1977. Did Sir Cyril Burt fake his research on heritability of intelligence? Part I. *Phi Delta Kappan* 58: 469–70.

Goldberger, A. S. 1972. Structural equation models in the social sciences. *Econometrica* 40: 979–1001.

————. 1976a. Jensen on Burks. *Educational Psychologist* 12: 64–78.

————. 1976b. Mysteries of the meritocracy. In *The IQ controversy*, ed. N. J. Block and G. Dworkin, pp. 265–79. New York: Pantheon Books.

————. 1976c. On Jensen's method for twins. *Educational Psychologist* 12: 79–82.

———. 1977. Twin methods: A skeptical view. In *Kinometrics: Determinants of socioeconomic success within and between families,* ed. P. Taubman, pp. 299–324. Amsterdam: North-Holland.

———. 1978a. Heritability. Manuscript, Social Systems Research Institute, University of Wisconsin, Madison. (Revised version of Newmarch Lectures in Economic Statistics at University College London, June 6 and 7, 1978.)

———. 1978b. Models and methods in the IQ debate: Part I. Revised. Manuscript, Social Systems Research Institute, University of Wisconsin, Madison.

———. 1978c. Pitfalls in the resolution of IQ inheritance. In *Genetic epidemiology,* ed. N. E. Morton and C. S. Chung, pp. 195–215. New York: Academic Press.

———. 1978d. The nonresolution of IQ inheritance by path analysis. *American Journal of Human Genetics* 30: 442–45.

Goldberger, A. S., and Duncan, O. D. eds. 1973. *Structural equation models in the social sciences.* New York: Seminar Press.

Goldberger, A. S., and Lewontin, R. C. 1976. Jensen's twin fantasy. Manuscript, Institute for Research on Poverty, University of Wisconsin, Madison.

Goldfeld, S. M., and Quandt, R. E. 1972. *Nonlinear methods in econometrics.* Amsterdam: North-Holland.

Gordon, R. A., and Rudert, E. E. 1979. Bad news concerning IQ tests. *Sociology of Education* 52: 174–90.

Gottesman, I. I. 1963. Genetic aspects of intelligent behavior. In *Handbook of mental deficiency,* ed. N. R. Ellis, pp. 253–96. New York: McGraw-Hill.

Gottesman, I. I., and Heston, L. L. 1972. Human behavioral adaptations: Speculations on their genesis. In *Genetics, environment, and behavior: Implications for educational policy,* ed. L. Ehrman, G. S. Omenn, and E. Caspari, pp. 105–22. New York: Academic Press.

Graybill, F. A. 1961. *An introduction to linear statistical models.* Vol. 1. New York: McGraw-Hill.

Guilford, J. P. 1968. The structure of intelligence. In *Handbook of measurement and assessment in the behavioral sciences,* ed. D. K. Whitla, pp. 215–59. Boston: Addison-Wesley.

Haggard, E. A. 1958. *Intraclass correlation and the analysis of variance.* New York: Dryden Press.

Hall, W. S., and Freedle, R. O. 1973. Towards the identification of cognitive operations in standard and nonstandard English usage. Princeton, N.J.: Educational Testing Service Research Bulletin 73.

Hall, W. S., and Guthrie, L. F. 1979. On the dialect question and reading. Champaign, Ill.: Center for the Study of Reading Technical Report 121.

Hall, W. S., and Tirre, W. C. 1979. The comparative environment of young children: Social class, ethnic, and situational differences. Champaign, Ill.: Center for the Study of Reading Technical Report 125.

Haller, M. H. 1963. *Eugenics: Hereditarian attitudes in American thought.* New Brunswick: Rutgers University Press.

Hamblin, R. L.; Buckholdt, D.; and Doss, H. 1970. Compensatory education: A new perspective. *Toledo Law Review,* pp. 459–99.

Harrison, D. S., and Trabasso, T. 1976. *Black English: A seminar.* Hillsdale, N.J.: Erlbaum Associates.

Hart, H. 1924. Correlations between intelligence quotients of siblings. *School and Society* 20: 382.

Hauser, R. M., and Goldberger, A. S. 1971. The treatment of unobservable variables in path analysis. In *Sociological methodology 1971,* ed. H. L. Costner, pp. 81–117. San Francisco: Jossey-Bass.

Hearnshaw, L. S. 1979. *Cyril Burt, Psychologist.* Ithaca: Cornell University Press.

Hecaean, H. 1962. Clinical symptomatology in right and left hemispheric lesions. In *Interspheric relations and cerebral dominance,* ed. V. B. Mountcastle, pp. 215–43. Baltimore: The Johns Hopkins Press.

Heise, D. R. 1969a. Problems in path analysis and causal inference. In *Sociological methodology 1969,* ed. E. F. Borgatta, pp. 38–73. San Francisco: Jossey-Bass.

―――. 1969b. Separating reliability and stability in test-retest correlation. *American Sociological Review* 34: 93–101.

Henderson, N. D. 1970. Genetic influences on the behavior of mice can be obscured by laboratory rearing. *Journal of Comparative Physiological Psychology* 72: 505–11.

Herman, L., and Hogben, L. 1933. The intellectual resemblance of twins. *Proceedings of the Royal Society of Edinburgh* 53: 105–29.

Herrnstein, R. 1971. I.Q. *Atlantic* 288: 43–64.

―――. 1973. *I.Q. in the meritocracy.* Boston: Atlantic–Little, Brown.

Hildreth, G. 1925. The resemblance of siblings in intelligence and achievement. Contributions to Education, no. 186. New York: Teacher's College.

Hirsch, J. 1971. Behavior-genetic analysis and its biosocial consequences. In *Intelligence: Genetic and environmental influences,* ed. R. Cancro, pp. 88–106. New York: Grune and Stratton.

Hogarth, R. M. 1974. Monozygotic and dyzygotic twins reared together: Sensitivity of heritability estimates. *British Journal of Mathematical and Statistical Psychology* 27: 1–13.

Holzinger, K. J. 1929. The relative effect of nature and nurture influences on twin differences. *Journal of Educational Psychology* 20: 241–48.

Horn, J. M.; Plomin, R.; and Rosenman, R. 1976. Heritability of personality traits in adult male twins. *Behavior Genetics* 6: 17–30.

Hornung, C. A. 1977. Social status, status inconsistency, and psychological stress. *American Sociological Review* 42: 623–38.

Huntley, R. M. C. 1966. Heritability of intelligence. In *Genetic and environmental factors in human ability,* ed. J. E. Meade and A. S. Parkes, pp. 201–18. Edinburgh: Oliver and Boyd.

Husen, T. 1953. *Tvilling studier.* Stockholm: Almqvist and Wiksell.

Jencks, C., and Brown, M. 1977. Genes and social stratification: A methodological exploration with illustrative data. In *Kinometrics: Determinants of socioeconomic success within and between families,* ed. P. Taubman, pp. 169–233. Amsterdam: North-Holland.

Jencks, C.; Smith, M.; Ackland, H.; Bane, M. J.; Cohen, D.; Gintis, H.; and Heyns, B. 1972. *Inequality: A reassessment of the effect of family and schooling in America.* New York: Basic Books.

Jensen, A. R. 1967. Estimation of the limits of heritability of traits by comparison of monozygotic and dizygotic twins. *Proceedings of the National Academy of Sciences* 58: 149–56.

———. 1969. How much can we boost IQ and scholastic achievement? *Harvard Educational Review* 39: 1–123.

———. 1970a. Can we and should we study race differences? In *Disadvantaged child: Compensatory education, a national debate,* ed. J. Hellmuth, pp. 124–57. New York: Brunner-Mazel.

———. 1970b. IQs of identical twins reared apart. *Behavior Genetics* 1: 133–46.

———. 1971. A note on why genetic correlations are not squared. *Psychological Bulletin* 75: 223–24.

———. 1972a. The case for IQ tests: Reply to McClelland. *Humanist* 32: 14.

———. 1972b. *Genetics and education.* New York: Harper and Row, 1972.

———. 1973. *Educability and group differences.* New York: Harper and Row.

———. 1974. Kinship correlations reported by Cyril Burt. *Behavior Genetics* 4: 1–28.

———. 1975. The meaning of heritability in the behavioral sciences. *Educational Psychologist* 11: 171–83.

———. 1976. The problem of genotype-environment correlation in the estimation of heritability from monozygotic and dizygotic twins. *Acta Geneticae Medicae et Gemellologiae* 25: 86–99.

———. 1977. Did Sir Cyril Burt fake his research on heritability of intelligence? Part II. *Phi Delta Kappan* 58: 471, 492.

———. 1978a. Sir Cyril Burt in perspective. *American Psychologist* 33: 499–503.

———. 1978b. The current status of the IQ controversy. *Australian Psychologist* 13: 7–27.

Jinks, J. L., and Eaves, L. J. 1974. IQ and inequality. *Nature* 248: 287–89.

Jinks, J. L., and Fulker, D. W. 1970. Comparison of the biometrical genetical, MAVA, and classical approaches to the analysis of human behavior. *Psychological Bulletin* 73: 311–49.

Jones, H. E. 1954. The environment and mental development. In *Manual of child psychology,* ed. L. Carmichael, pp. 631–96. New York: Wiley.

Jordan, W. D. 1974. *The white man's burden: Historical origins of racism in the United States.* New York: Oxford University Press.

Jöreskog, K. G. 1973. A general method for estimating a linear structural equation system. In *Structural equation models in the social sciences,* ed. A. S. Goldberger and O. D. Duncan, pp. 85–112. New York: Seminar Press.

Juel-Nielsen, N. 1965. Individual and environment: A psychiatric-psychological investigation of monozygous twins reared apart. *Acta Psychiatrica et Neurologica Scandinavica*, Monograph supp. 183.

Julesz, B. 1972. Talk given before the Department of Psychology, University of California, San Diego.

Kagan, J. 1964. Information processing in the child: significance of analytic and reflective attitudes. *Psychological Monographs* 78: whole no. 578.

Kagan, J.; Moss, H. A.; and Siegal, I. E. 1963. Psychological significance of styles of conceptualization. In *Basic cognitive process in children*. Society for Research in Child Development, monograph 86. Chicago: Univ. of Chicago Press.

Kamin, L. J. 1973. Heredity, intelligence, politics, and psychology. Talk delivered at Eastern Psychological Association meeting, New York, N.Y.

———. 1974. *The science and politics of I.Q.* Potomac, Md.: Erlbaum Associates.

———. 1975. Review essay on A. R. Jensen, *Educability and group differences* and A. R. Jensen, *Genetics and education. Contemporary Psychology* 20: 545–47.

———. 1978. Transfusion syndrome and the heritability of IQ. *Annals of Human Genetics*, London 42: 161–71.

Kendall, M. G. 1960. *The advanced theory of statistics.* 3rd ed. Vol. 1. New York: Hafner.

Kerlinger, F. N. 1973. *Foundations of behavioral research.* 2d ed. New York: Holt, Rinehart and Winston.

Kerlinger, F. N., and Pedhazur, E. J. 1973. *Multiple regression in behavioral research.* New York: Holt, Rinehart and Winston.

Kimura, D. 1963. Right temporal-lobe damage. *Archives of Neurology* 8: 264–71.

———. 1964. Left-right differences in the perception of melodies. *Quarterly Journal of Experimental Psychology* 15: 166–71.

———. 1973. The asymmetry of the human brain. *Scientific American* 228: 70–73.

Koch, H. L. 1966. *Twins and twin relations.* Chicago: Univ. of Chicago Press.

Kogan, N. 1973. Creativity and cognitive style: A life-span perspective. In *Life-span developmental psychology,* ed. P. B. Baltes and K. W. Schaie, pp. 145–78. New York: Academic Press.

Land, K. C. 1969. Principles of path analysis. In *Sociological methodology 1969,* ed. E. F. Borgatta, pp. 3–37. San Francisco: Jossey-Bass.

Layzer, D. 1974. Heritability analysis of IQ scores: Science or numerology? *Science* 183: 1259–66.

———. 1975. Review essay on L. J. Kamin, *The science and politics of IQ. Scientific American* 230: 126–28.

Leahy, A. M. 1932. A study of certain selective factors influencing prediction of the mental status of adopted children. *Pedagogical Seminary and Journal of Genetic Psychology* 41: 294–329.

————. 1935. Nature-nurture and intelligence. *Genetic Psychology Mono-graphs* 17: 236–308.

Lee, D. 1968. Codifications of reality: Lineal and nonlineal. In *Every man his way,* ed. A. Dundes, pp. 329–43. Englewood Cliffs, N.J.: Prentice-Hall.

Levy, J. 1970. Information processing and higher psychological functions in the disconnected hemispheres of human commissurotomy patients. Ph.D. dissertation, California Institute of Technology.

Lewontin, R. C. 1970. Race and intelligence. *Bulletin of the Atomic Scientists* 26: 2–8.

————. 1974. The analysis of variance and the analysis of causes. *American Journal of Human Genetics* 26: 400–11.

————. 1975. Genetic aspects of intelligence. *Annual Review of Genetics* 9: 387–405.

Light, R. J., and Smith, P. V. 1969. Social allocation models of intelligence: A methodological inquiry. *Harvard Educational Review* 39: 484–510.

Loehlin, J. C. 1978. Heredity-environment analyses of Jencks's IQ correlations. *Behavior Genetics* 8: 415–36.

Loehlin, J. C.; Lindzey, G.; and Spuhler, J. N. 1975. *Race differences in intelligence.* San Francisco: W. H. Freeman and Co.

Loehlin, J. C.; Vandenberg, S. G.; and Osborn, R. T. 1973. Blood-group genes and negro-white ability differences. *Behavior Genetics* 3: 263–70.

Luria, A. R. 1976. *Cognitive development: Its cultural and social foundations.* Cambridge: Harvard University Press.

Lush, J. L. 1949. Heritability of quantitative characters in farm animals. *Heredi-tas,* Supp. vol.: 356–75.

Madsen, I. N. 1924. Some results with the Stanford revision of the Binet-Simon tests. *School and Society* 19: 559–62.

Marsh, F. J.; Bogen, J. E.; and TenHouten, W. D. 1971. Are there two kinds of thinking? Final Report, pt. 1. O.E.O. study 500-5135, University of California, Riverside.

Mazur, A., and Robertson, L. S. 1972. *Biology and social behavior.* New York: Free Press.

McCall, R. B.; Appelbaum, M. I.; and Hogarty, P. S. 1973. Developmental changes in mental performance. *Monographs of the Society for Research in Child Development* 38: 1–84.

McClearn, G. E., and DeFries, J. C. 1973. *Introduction to behavior genetics.* San Francisco: W. H. Freeman and Co.

McClelland, D. C. 1967. Wanted: A new self-image for women. In *The woman in America,* ed. R. J. Lifton, pp. 173–92. Boston: Beacon.

McNemar, Q. 1938. Special review, Newman, Freeman, and Holzinger's twins: A study of heredity and environment. *Psychological Bulletin* 35: 237–49.

————. 1942. *The revision of the Stanford-Binet scale: An analysis of the stand-ardization data.* Boston: Houghton-Mifflin.

Milner, B. 1971. Interhemispheric differences in the localization of psycholog-ical processes in man. *British Medical Bulletin* 27: 272–77.

Morton, N. E. 1972. Human behavioral genetics. In *Genetics, environment, and behavior: Implications for educational policy,* ed. L. Ehrman, G. S. Omenn, and E. Caspari, pp. 247–65. New York: Academic Press.

———. 1974. Analysis of family resemblance I. Introduction. *American Journal of Human Genetics* 26: 318–30.

———. 1978. Comment. In *Genetic epidemiology,* ed. N. E. Morton and C. S. Chung, p. 193. New York: Academic Press.

Morton, N. E., and Chung, C. S., eds. 1978. *Genetic epidemiology.* New York: Academic Press.

Morton, N. E., and Rao, D. C. 1978. Quantitative inheritance in man. *Yearbook of Physical Anthropology* 21: 12–41.

Muller, H. J. 1925. Mental traits and heredity. *Journal of Heredity* 16: 433–48.

Munsinger, H. 1977. The identical-twin transfusion syndrome: A source of error in estimating IQ resemblance and heritability. *Annals of Human Genetics* 40: 307–21.

Namboodiri, N. K.; Carter, L. F.; and Blalock, H. M., Jr. 1975. *Applied multivariate analysis and experimental designs.* New York: McGraw-Hill.

Nebes, R. D. 1972. Dominance of the minor hemisphere in commissurotomized man on a test of figural unification. *Brain* 94: 633–38.

Newman, H. H.; Freeman, F. N.; and Holzinger, K. J. 1937. *Twins: A study of heredity and environment.* Chicago: Univ. of Chicago Press.

Nichols, P. L. 1970. The effects of heredity and environment on intelligence test performance in 4 and 7 year white and negro sibling pairs. Ph.D. dissertation, University of Minnesota. Ann Arbor: University Microfilms, no. 71-18874.

Nichols, P. L., and Broman, S. H. 1973. Familial factors associated with IQ at 4 years. Paper presented at annual meeting, Behavior Genetics Association, Chapel Hill, N.C.

Nichols, R. C. 1965. The National Merit twin study. In *Methods and goals in human behavior genetics,* ed. S. G. Vandenberg, pp. 231–43. New York: Academic Press.

Olneck, M. R. 1977. On the use of sibling data to estimate the effects of family background, cognitive skills, and schooling. In *Kinometrics: Determinants of socioeconomic success within and between families,* ed. P. Taubman, pp. 125–62. Amsterdam: North-Holland.

Ornstein, R. E., ed. 1973. *The nature of human consciousness.* San Francisco: W. H. Freeman and Co.

Osborn, R. T., and Gregor, A. J. 1968. Racial differences in heritability estimates for tests of spatial ability. *Perceptual and Motor Skills* 27: 735–39.

Osborn, R. T., and Miele, F. 1969. Racial differences in environmental influences on numerical ability as determined by heritability estimates. *Perceptual and Motor Skills* 28: 535–38.

Osofsky, G. 1971. *Harlem: The making of a ghetto.* 2d ed. New York: Harper Torchbooks.

Outhit, M. C. 1933. A study of the resemblance of parents and children in general intelligence. *Archives of Psychology* 149: 1–60.

Partanen, J.; Bruun, K.; and Markkanen, T. 1966. *Inheritance of drinking behavior: A study of intelligence, personality, and use of alcohol of adult twins.* Helsinki: The Finnish Foundation for Alcohol Studies.

Patterson, K., and Bradshaw, J. L. 1975. Differential hemispheric mediation of nonverbal visual stimuli. *Journal of Experimental Psychology (Human Perception)* 1: 246–52.

Pintner, R. 1918. The mental indices of siblings. *Psychological Review* 25: 252–55.

Plomin, R.; DeFries, J. C.; and Loehlin, J. C. 1977. Genotype-environment interaction and correlation in the analysis of human behavior. *Psychological Bulletin* 84: 309–22.

Powell, M. J. D. 1971. Recent advances in unconstrained optimization. *Mathematical Programming* 1: 26–57.

Ramirez, M., III, and Castañeda, A. 1974. *Cultural democracy, bicognitive development, and education.* New York: Academic Press.

Rao, D. C., and Morton, N. E. 1974. Path analysis of family resemblance in the presence of gene-environment interaction. *American Journal of Human Genetics* 26: 767–72.

———. 1978. IQ as a paradigm in genetic epidemiology. In *Genetic epidemiology,* ed. N. E. Morton and C. S. Chung, pp. 145–82. New York: Academic Press.

Rao, D. C.; Morton, N. E.; and Yee, S. 1974. Analysis of family resemblance II. A linear model for familial correlation. *American Journal of Human Genetics* 26: 331–59.

———. 1976. Resolution of cultural and biological inheritance by path analysis. *American Journal of Human Genetics* 28: 228–42.

———. 1978. Resolution of cultural and biological inheritance by path analysis: Corrigenda and reply to Goldberger letter. *American Journal of Human Genetics* 30: 445–48.

Rodriguez, E. 1974. Inside ETS—Or the plot to multiple-choice us from cradle to grave. *Washington Monthly* (March): 5–12.

Rubin, D. B. 1979. Letter: Burt's tables. *Science* 204: 244–46.

Samples, R. E. 1975. Learning with the whole brain. *Human Behavior* (February): 16–23.

Samuda, R. J. 1975a. *Psychological testing of American minorities: Issues and consequences.* New York: Dodd, Mead.

———. 1975b. From ethnocentrism to a multi-cultural perspective. *Journal of Afro-American Issues* 3: 4–18.

Sanday, P. R. 1972. On the causes of IQ differences between groups with implications for social policy. *Human Organization* 31: 411–24.

Saudek, R. 1934. A British pair of identical twins reared apart. *Character and Personality* 3: 17–39.

Scarr, S. 1968. Environmental bias in twin studies. *Eugenics Quarterly* 15: 34–40.

―――. 1969. Effects of birth weight on later intelligence. *Social Biology* 16: 249–56.

Scarr, S., and Weinberg, R. A. 1976. IQ test performance of black children adopted by white families. *American Psychologist* 31: 726–39.

―――. 1977. Intellectual similarities within families of both adopted and biological children. *Intelligence* 1: 170–91.

―――. 1978. The influence of "family background" on intellectual attainment. *American Sociological Review* 43: 674–92.

Scarr, S.; Pakstis, A. J.; Katz, S. H.; and Barker, W. B. 1977. Absence of a relationship between degree of white ancestry and intellectual skills within a black population. *Human Genetics* 39: 69–86.

Scarr-Salapatek, S. 1971a. Race, Social Class, and IQ. *Science* 174: 1285–95.

―――. 1971b. Unknowns in the IQ equation. *Science* 174: 1223–28.

―――. 1974. Some myths about heritability and IQ. *Nature* 251: 463–64.

―――. n.d. Twin method: Defense of a critical assumption. Manuscript.

Scarr-Salapatek, S., and Weinberg, R. A. 1975. When black children grow up in white homes. *Psychology Today* 9: 80–82.

Schwartz, M., and Schwartz, J. 1974. Evidence against a genetical component to performance on IQ tests. *Nature* 248: 84–85.

―――. 1975. Statistical methods in the experimental analysis of the heritability of IQ performance. Manuscript.

―――. 1976. Comment on "IQs of identical twins reared apart." *Behavior Genetics* 6: 367–68.

Segall, M. H. 1976. *Human behavior and public policy: A political psychology.* Elmsford, N.Y.: Pergamon Press.

Shields, J. 1954. Personality differences in neurotic traits in normal twin schoolchildren. *Eugenics Review* 45: 213–47.

―――. 1962. *Monozygotic twins brought up apart and brought up together.* London: Oxford University Press.

Shockley, W. 1969. A polymolecular interpretation of growth rates of social problems. Paper read before the National Academy of Sciences, Washington, D.C.

―――. 1970. New methodology to reduce the environment-heredity uncertainty about dysgenics. Paper read before the National Academy of Sciences, Rice University, Houston, Texas.

―――. 1972. The apple-of-God's-eye obsession. *Humanist* 32: 16–17.

―――. 1973. Forum letter. Foundation for Research and Education on Eugenics and Dysgenics (FREED), April 23.

―――. 1978. Humanitarianism gone berserk? *Dallas Times Herald,* September 17.

Shuey, A. 1966. *The testing of negro intelligence.* 2d ed. New York: Social Science Press.

Skodak, M. 1950. Mental growth of adopted children in the same family. *Pedagogical Seminary and Journal of Genetic Psychology* 77: 3–9.

Skodak, M., and Skeels, H. M. 1949. A final follow-up study of 100 adopted children. *Pedagogical Seminary and Journal of Genetic Psychology* 75: 85–125.

Smith, A. 1974. Dominant and nondominant hemispherectomy. In *Hemispheric disconnection and cerebral function*, ed. M. Kinsbourne and W. L. Smith, pp. 5–33. Springfield, Ill.: Charles C. Thomas.

Smith, K. W., and Sasaki, M. S. 1979. Decreasing multicollinearity: A method for models with multiplicative functions. *Sociological Methods and Research* 8: 35–56.

Smith, R. T. 1965. A comparison of socioenvironmental factors in monozygotic and dizygotic twins, testing an assumption. In *Methods and goals in human behavior genetics*, ed. S. G. Vandenberg, pp. 45–61. New York: Academic Press.

Snider, B. 1955. A comparative study of achievement test scores of fraternal and identical twins and siblings. Ph.D. dissertation, State University of Iowa.

Snygg, D. 1938. The relation between the intelligence of mothers and their children living in foster homes. *Journal of Genetic Psychology* 52: 401–6.

Southwood, K. E. 1978. Substantive theory and statistical interaction: Five models. *American Journal of Sociology* 83: 1154–1203.

Sperry, R. W. 1968. Mental unity following surgical disconnection of the cerebral hemisphere. In *The Harvey lectures*, ser. 62, pp. 293–323. New York: Academic Press.

———. 1975. Left-brain, right-brain. *Saturday Review* 58 (August 9): 30–33.

Sperry, R. W.; Gazzaniga, M. S.; and Bogen, J. E. 1969. Interhemisphere relationships: The neocortical commissures: Syndromes of hemisphere disconnection. In *Handbook of clinical neurology*, vol. 4, ed. P. J. Vinken and G. W. Bruyn, pp. 273–90. Amsterdam: North-Holland.

Spielman, W., and Burt, C. 1926. In *A study in vocational guidance*, ed. F. Gaw, L. Ramsey, M. Smith, and W. Spielman, pp. 12–17. Report no. 33. London: H. M. Stationery Office.

Spuhler, J. N., and Lindzey, G. 1967. Racial differences in behavior. In *Behavior-genetic analysis*, ed. J. Hirsch, pp. 366–414. New York: McGraw-Hill.

Stigler, S. M. 1979. Letter: Burt's tables. *Science* 204: 242–44.

Stinchcombe, A. 1972. The social determinants of success. *Science* 178: 603–4.

Stocks, P., and Karn, M. N. 1933. A biometric investigation of twins and their brothers and sisters. *Annals of Eugenics* 5: 1–55.

Stone, C. 1975. Standardized tests: True or false. *Black Collegian* 6.

Tanser, H. A. 1941. Intelligence of negroes of mixed blood in Canada. *Journal of Negro Education* 10: 650–52.

Taubman, P. 1976. The determinants of earning: Genetics, family and other environments. *American Economic Review* 66: 858–70.

————. ed. 1977. *Kinometrics: Determinants of socioeconomic success within and between families.* Amsterdam: North-Holland.

Taylor, H. F. 1973a. Linear models of consistency: Some extensions of Blalock's strategy. *American Journal of Sociology* 78: 1192–1215.

————. 1973b. Playing the dozens with path analysis: Methodological pitfalls in Jencks et al., "Inequality." *Sociology of Education* 46: 433–50.

————. 1974. Review essay, R. Herrnstein, *IQ in the meritocracy. Journal of the American Statistical Association* 69: 1043–44.

————. 1975. The econometrics of cognitive consistency: Some comments on a general model. Paper read at the annual meeting of the American Sociological Association, Social Psychology Section, San Francisco.

————. 1977. The equations of oppression. In *Black separatism and social reality,* ed. R. L. Hall, pp. 233–42. New York: Pergamon Press.

Taylor, H. F., and Hornung, C. A. 1979. On a general model for social and cognitive consistency. *Sociological Methods and Research* 7: 259–87.

TenHouten, W. D. 1976. More on split-brain research, culture and cognition. *Current Anthropology* 17: 503–6.

————. 1977. More on split-brain research and anthropology. *Current Anthropology* 18: 344–46.

TenHouten, W. D., and Kaplan, C. D. 1973. *Science and its mirror image.* New York: Harper and Row.

Tumin, M. M., ed. 1963. *Race and intelligence.* New York: Anti-Defamation League of B'nai B'rith.

Vandenberg, S. G. 1962. The heredity abilities study: Heredity components in a psychological test battery. *American Journal of Human Genetics* 14: 220–37.

————. 1967. Hereditary factors in psychological variables in man, with special emphasis on cognition. In *Genetic diversity and human behavior,* ed. J. N. Spuhler, pp. 99–133. New York: Viking Fund Publications in Anthropology, no. 45.

————. 1968. The nature and nurture of intelligence. In *Biology and Behavior: Genetics,* ed. D. Glass, pp. 3–58. New York: Rockefeller Univ. Press.

————. 1970. A comparison of heritability estimates of U.S. negro and white high school students. *Acta Geneticae Medicae et Gemellologiae* 19: 280–84.

————. 1971. What do we know today about the inheritance of intelligence and how do we know it? In *Intelligence: Genetic and environmental influences,* ed. R. Cancro, pp. 182–218. New York: Grune and Stratton.

Vandenberg, S. G., and Johnson, R. C. 1968. Further evidence of the relation between age of separation and similarity in IQ among pairs of separated identical twins. In *Progress in human behavior genetics,* ed. S. G. Vandenberg, pp. 215–19. Baltimore: The Johns Hopkins Press.

Vasgird, D. R. 1975. Oh God, oh Galton, Mr. Herrnstein! *Crisis* 82 (November): 341–47.

Vogel, F.; Schalt, E.; Kriger, J.; Propping, P.; and Lehnert, K. F. 1979. The electroencephalogram (EEG) as a research tool in human behavior genetics: Psychological examination in healthy males with various inherited EEG variants. *Human Genetics* 47: 1–45.

Wade, N. 1976. IQ and heredity: Suspicion of fraud beclouds classic experiment. *Science* 194: 916–19.

————. 1978. California court is forum for latest round in IQ debate. *Science* 201: 1106–8.

Watson, J. B. 1930. *Behaviorism*. Rev. ed. Chicago: Univ. of Chicago Press.

Werts, C. E., and Linn, R. L. 1970. Path analysis: Psychological examples. *Psychological Bulletin* 74: 193–212.

Wictorin, M. 1952. *Bidrag till räknefärdighetens psykologi, en tvillingsunderökning*. Göteborg: Elanders Boktryckeri.

Willerman, L., and Churchill, J. A. 1967. Intelligence and birth weight in identical twins. *Child Development* 38: 623–29.

Williams, R. L. 1975. The BITCH-100: A culture-specific test. *Journal of Afro-American Issues* 3: 103–16.

Willoughby, R. R. 1928. Family similarities in mental-test abilities. In *Twenty-seventh yearbook of the national society for the study of education*, pt. 1, pp. 55–59. Bloomington: Public School Publishing Co.

Wilson, P. T. 1934. A study of twins with special reference to heredity as a factor determining differences in environment. *Human Biology* 6: 324–54.

Witkin, H. A. 1967. A cognitive style approach to cross-cultural research. *International Journal of Psychology* 4: 233–50.

Witkin, H. A., and Goodenough, D. R. 1976. Field dependence revisited. Princeton, N.J.: Educational Testing Service Research Bulletin 76–39.

Witkin, H. A.; Dyk, R. B.; Faterson, H. F.; Goodenough, D. R.; and Karp, S. A. 1962. *Psychological differentiation: Studies of development*. New York: Wiley.

Witkin, H. A.; Lewis, H. B.; Hertzman, M.; McHover, K.; Meissner, P. B.; and Wapner, S. 1954. *Personality through perception*. New York: Harper.

Witkin, H. A.; Moore, C. A.; Oltman, P. K.; Goodenough, D. R.; Friedman, F.; Owen, D. R.; and Raskin, E. 1977. Role of the field-dependent and field-independent cognitive styles in academic evolution: A longitudinal study. *Journal of Educational Psychology* 69: 197–211.

Wright, S. 1921. Correlation and causation. *Journal of Agricultural Research* 20: 557–85.

————. 1925. Corn and hog correlations. Department Bulletin no. 1,300, U.S. Department of Agriculture, U.S. Government Printing Office.

————. 1931. Statistical methods in biology. Papers and Proceedings of the 92d Annual Meeting. *Journal of the American Statistical Association Supplement* 26: 155–63.

————. 1934. The method of path coefficients. *Annals of Mathematical Statistics* 5: 161–215.

Yin, R. K. 1969. Looking at upside-down faces. *Journal of Experimental Psychology* 81: 141–45.

————. 1970. Face recognition by brain-injured patients: A dissociable ability? *Neuropsychologia* 8: 395–402.

Yule, G. U., and Kendall, M. G. 1950. *An introduction to the theory of statistics.* New York: Hafner.

Zajonc, R. B., and Markus, G. B. 1975. Birth order and intellectual development. *Psychological Review* 82: 74–88.

Zazzo, R. 1960. *Les jumeax, le couple et la personne.* Vol. 2. Paris: Presses Universitaires de France.

Index

(Page numbers in italics designate tables.)

A

Additivity, *28*, 71, 114, 115, 121, 122, 134, 137, 166, 212
  Honolulu models and, 175, 180
  hypothesis of Jinks and Fulker, 159–160
  interaction question and, 155, 156
  See also Genes; Nonadditive path model illustration
Admissible solution, 174
Adopted-adopted pairs reared together, *48*, 136, 152, 153, 204, 207
Adopted-natural pairs reared together, *48*, 136, 152, 153, 204, 207
Adoptions, 177
Adult IQ environmentability, 175–176, 187–192
Adult IQ heritability, 175–176, 187–192, 205
African blood-group ancestry. *See* Blood-group differences (African)
Age standardization of tests, 80, 81, 83, 101, 102, 113, 131
Alleles. *See* Genes, interaction of forms of (alleles)
Althauser, R. P., 162
Amino acids, genetic codes and, 27
Analysis of variance (ANOVA), 15, 238
Anthroposcopy of MZ twin pairs, 79
Arithmetic errors, 81, 118, 153, 174, 212–213. *See also* Measurement errors
Army Alpha scores, 39, *47*
Army Beta scores, 39

Artificial inflation of IQ correlation, 22, 29, 54, 80, 83, 101, 110, 115, 141–142, 152, 206
  inflation quantity and, 142–143, *144–145, 146*, 147
Assortative mating, 16, *28*, 152, 180, 203
Attenuation correction. *See* Correction for attenuation

B

Babson, S. G., 102
Baldwin, A. L., 223
Bandura, A., 220
Bayley Mental Scale IQ test, 66
Behavioral phenotypes, 11. *See also* Phenotypes
Behrman, J., 6, 150, 213
Benton, A. L., 224
Beta weights, defined, 120
Between-class variance, 17
Between-group differences in IQ heritability, 9
  race difference and, 64–68
Between-group heritability, 64
Bias. *See* Sample bias
Binet IQ, *46, 47*
  "London," Jencks and, *48*, 53
  See also Stanford-Binet IQ test
Birmingham group, 6, 175, 186, 205
  kinship data and, 29
  models of, 202–204
  See also Eaves, L.; Jinks, J. L.

Birth order, 127–128
Birth weight
    asymmetry mechanism and, 128–129
    effect of, on IQ, 104, 109
Blacks. See Race
Blalock, H. M., Jr., 17, 119, 133, 162
Blewett, D. B., 46, 59
Block, N. J., 243
Blood-group differences (African), 67–68, 209
Blood grouping of MZ twin pairs, 79, 84
Bogen, J. E., 219, 223, 224
Bogen, G. M., 223
Bogus heritability. See IQ heritability, bogus, kinship correlation and
Bogus indexes, 150, 167
Bohrnstedt, G. W., 162
Boring, E. G., 215
Bouchard, T., 216
Boudon, R., 119
Brain
    organization of, genetic differences and, 219
    split-brain theory and, 223–224
Broad heritability, 15, 16, 121, 203, 229. See also Narrow heritability
Broman, S. H., 127
Bronfenbrenner, U., 111
Brown, M., 127, 150, 213
Burks, B. S., 47, 107, 118, 175, 177, 187, 200, 236
    culture index of, 186
Burt, Cyril, 5, 12, 14, 15, 16, 72, 121, 140, 204, 213, 235, 241
    corrected for unreliability, 55
    Jensen's kinship equation and, 45, 46, 47, 51–52, 53, 54–55, 56, 57, 59
    kinship data of, 29, 31, 32, 34
    scandal concerning, 34–41
    separated MZ twins study of, 20, 21, 76
    Stanford-Binet test and, 23–24

C

Carmon, A., 224
Casteñeda, A., 223
Cavalli-Sforza, L. L., 239
Ceteris paribus (for environment), 113–114

Child IQ environmentability, 175–176, 187–192, 208
Child IQ heritability, 175–176, 187–192, 204, 205, 208
Chronological age, 102
Chung, C. S., 175
Churchill, J. A., 102
Clone, MZ twins as, 18
Coefficient of heritability (h^2). See IQ heritability, coefficient of heritability (h^2) and
Cognitive abilities, 132
Cognitive process, heritability of, 132
Cognitive style
    analytic, 220
    integrative, 220, 221
    Vasgird's overview of, 218–225
Cohen, D. J., 61
Cohen, J., 35, 38
Cohen, R. A., 131, 162, 218, 219, 221–222
Collateral kinship model, 124–137
Collateral kinships, 28
    identical (MZ) twins and, 11
Collateral pairs, 177, 180
Color blindness of MZ twin pairs, 79
Composite scores (Dominoes and Mill Hill), 236. See also Dominoes test; Raven's Mill Hill Vocabulary Scale
Compound paths. See Path analysis of IQ heritability, compound paths and
Conrad, H. S., 46, 213
Constructs (inferential variable), 11, 177, 233–234
    IQ score "index" and, 215
Conway, J., 38
Correction for attenuation, 22, 53, 54, 132, 133
    IQ scores and, 113
    Jensen's kinship correlation and, 28
    residual variance and, 115
Costner, H. L., 133, 233
Covariance. See Gene-environment covariance
Creativity, cognitive styles and, 219
CRM, Society Today, 29, 110
Cromie, W. J., 219

Cross-pair gene-environment correlation. *See* Gene-environment correlation, cross-pair
Cultural bias. *See* Sample bias, cultural
"Cultural transmission," 202
Culture index (Burks), 186
Cyril Burt scandal. *See* Burt, Cyril, scandal concerning

D

Dasen, P. R., 67
Data sets, Honolulu group. *See* Honolulu group, data sets of
DeFries, J. C., 13, 15, 22, 50, 86, 110, 121, 149, 158, 162, 233
  heritability and, 64–65, 66
  kinship data and, 29
Degree of freedom (df), 18
  Honolulu group and, *183, 185,* 187, 192
DeLemos, M. N., 67
DeRenzi, E., 224
Derived parameters (Honolulu group), *183, 185,* 188
Difference correlation, inflation of, 80, 110, 212
Dillard, J. L., 131
Dimensionality of IQ tests, cultural bias and, 131
Direct line kinships. *See* Kinships, direct line
Dixon, study of defecation behavior, 158
Dizygotic-fraternal twins. *See* Twins, DZ (dizygotic-fraternal)
Dobzhansky, T., 39, 71, 110, 159
Dominance (genetic), 16, *28,* 202–203. *See also* Genes, dominance and
Dominoes test, *20,* 79–80
  composite scores and, 236
Dorfman, D. D., Burt scandal and, 39–41, 73
Double-entry correlation, 17
Down's syndrome, 209
Duncan, O. D., 119, 122
  Honolulu group and, 177, 187, 201
Dworkin, G., 243

E

Eaves, L., 6, 50, 202, 203
  kinship data of, 29
Eckland, Bruce, 6, 76
Education, 1
  policies concerning, IQ heritability and, 3
  special programs in, IQ change and, 113
Educational Testing Service (ETS), 229
Eleftherious, study of avoidance behavior, 158
Environment, 2, 9, 13, 15, 156, 189, 201, 222, 240
  adult residual, 186
  ceteris paribus and, 113–114
  child-adult (Honolulu group), 187–192
  child residual, 186
  differences, linear function of, 161
  disturbing effects of, Burt scandal and, 36, 37, 38
  enriched, 64
  equal assumption of, 191
  genotype and, 14, 26, 116–117, 126–127
  importance of studying wide range of, 158
  intrapair (family), 52, 90, 104, 105, 125, 169, 211
  of MZ twins, 75
  noncommon, 176, 177
  postnatal, 134
  residual, 186, 190
  restrictions on, 21
  specific (noncommon), 128
  specific (nonfamily), 52, 105, 116, 125, 138, 169, 175, 198, 211
  as synonymous with "everything else," 115
  See also Adult IQ environmentability; Child IQ environmentability; Environmental similarity; Gene-environment correlation; Gene-environment covariance; Intra-uterine environment; Social environment
Environmentability (e^2) coefficient, 15, 121, 136

Environmental correlation, 125, 136, 138
  Honolulu group and, 190
  inflation quantity and, 147
Environmental index, 177, 181, 202, 203
Environmental manipulation, 13–14
Environmental similarity, 19, 74, 136, 141,
    168, 173, 216
  conclusions concerning (Taylor), 209–
    211
  gene-environment covariance and,
    68–72
  intrapair, 124–125, 167–168, 198, 214,
    232, 240
  IQ similarity and, 12
  Jensen's kinship equation and, 42, 60,
    61–63
    intrapair (family)–specific (nonfam-
    ily), 52
  Kamin's treatment of, 235
  kinship categories and, 26–27
  "minimal," 238
  of separated identical (MZ) twins, 75–
    77, 134–135
    artificial inflation of data and, 109–
    111
    family environment similarity and,
    90–92
    Jencks's path estimates and, 103–109
    Juel-Nielsen study and, 83–84, 95
    late separation and, 85–87
    Newman study and, 81–83, 94–95
    reunions and, 87–90
    Shields's study and, 78–81, 94
    social environment similarity and,
    92–101
  See also Environment; Social environ-
    ment
Environmental treatments, 72. See also
    Treatment effects
Environmental variables, 14, 69
  heritability estimation and, 115–116
Enzymes, genetic codes and, 27
Epistasis, 16
Erlenmeyer-Kimling, L., 72, 118, 241
  environment-genetic strains and, 158
  Jensen's kinship equation and, 43, 45,
    53, 55, 171
  kinship data and, 29, 30, 31, 32, 33, 140

Errors. See Arithmetic errors; Measure-
    ment errors; Specification errors
Eugenics, defined, 7
Exact identification (equation system),
    138, 172
Examiner. See Same-examiner effect
Exogenous variables, 153
Expected correlation (Honolulu group),
    186–187
Eysenck, H. J., 16, 46, 59, 67, 72, 87, 110,
    112, 137, 157, 203, 204, 206, 209,
    211
  Burt scandal and, 37, 39
  heritability estimation and, 115, 118,
    120, 121, 126
  Jencks's path estimates and, 103
  kinship data and, 29, 32
  race differences in intelligence and, 5–6
  separated identical (MZ) twin correla-
    tion and, 18–19, 21, 76, 129

F

Fabrication device (Burt), 40–41
Falconer, D. S., 135
  heritability index of, 42, 148, 167
Family environment. See Environment, in-
    trapair (family)
Family size, 127
Featherman, D. L., 177, 186, 187, 201
Feldman, N. W., 14, 65, 239
Fernandez, R., 110
"Final assessments" (Burt), 23–24, 35, 36,
    37, 52
Fingerprint pattern of MZ twins, 79, 81, 84
First cousins, 26, 28
Fisher, R. A., 14, 202
Foster parent and child, kinship studies
    and, 33
Freedle, R. O., 131
Freeman, F. N., 47, 82, 191, 200, 201
Free parameters (Honolulu group), 183,
    185, 187, 192, 197
F-test for ratio of variance, 17–18
Fulker, D. W., 6, 63, 71, 76, 161, 202, 212,
    238, 239 ·
  additivity hypothesis of, 159–160
  Jencks's path estimates and, 103

Fulker, D. W. (*continued*)
    kinship data of, 29
Fuller, J. L., 158
Fundamental equation in heritability, 129,
    134, 166

G

Galton, Sir Francis, 1, 7–8, 17, 119
Gardner, I. C., 236
Gazzaniga, M. S., 223
Gene-environment correlation, 105, 136,
    139, 152, 154, 166, 168, 193, 194,
    207, 214–215, 232–233, 240
    active, passive, and reactive, 69
    Birmingham group and, 202
    conclusions concerning (Taylor), 210–
    211
    cross-pair, 126, 134, 165, 169–170, 180
    heritability estimation and, 115, 118
    parental, 180
Gene-environment covariance, 13, 14, 68–
    72, 74, 151
    childhood, 176, 180
    heritability estimation and, 116
    Honolulu group and, 176, 180, *185*
    IQ variance and, 15
    Jencks's path estimates and, 103, 104,
    108
    Jensen's kinship equation and, 42, 172
    race difference and, 66
Gene-environment interaction, 1, 13, 14,
    55, 68–72, 74
    defined, 154–162
    dual character of, 159
    heritability estimation and, 114–115,
    116
    IQ variance and, 15
    nonadditive path model illustration
    and, 162–166
Generalized kinship equation, 41–42. *See
    also* Jensen, Arthur R., kinship
    equation of
Genes, 12, 166, 174
    additive, *28*
    dominance and, 16
    of DZ twins and, 25–26
    environment and, 14
    of identical (MZ) twins, 25–26, 75

Genes (*continued*)
    for intelligence, 12, 25, 69, 111, 114–115,
    119, 206
    interaction of forms of (alleles), 16
    IQ-occupation link and, 39
    IQ scores and, 4
    Jensen's kinship equation and, 43, 55
    shared, kinship data and, 31
Genetic correlation, 177
Genetic determinations, 12, 16, 42
Genetic differences, organization of the
    brain and, 219
Genetic dominance. *See* Dominance (ge-
    netic); Genes, dominance and
Genetic factors, 154, 168, 209, 214, 216, 232
    DZ-MZ twins and, 25–26
    gene-environment covariance, gene-
    environment interaction and,
    68–72
    inheritance of intelligence, coefficient
    of heritability and, 10–16
    in IQ heritability literature, 1–9
    Jencks's path estimates and, 103, 104,
    108
    Jensen's genetic strain (mice and rats)
    and, 157
    Jensen's kinship equation and, 42–43,
    61–63
    race differences and, 64–68
    test predictability, age and, 102
    See also Adult IQ heritability; Child IQ
    heritability; IQ heritability
Genetic heritability. *See* Adult IQ herita-
    bility; Child IQ heritability; IQ
    heritability
Genetic similarity, kinship categories and,
    26–27
Genotype, 10–11, 15, 70, 71, 154, 156, 231
    environments and, 14, 26, 116–117,
    126–127, 162
    hypothetical, measured phenotype and,
    13
    importance of studying wide range of,
    158
    "intelligence," 65–66, 109, 120, 159, 169,
    233
    mice-rats experiments and, 157
    multiplicative functions of, phenotypic
    variable and, 162

Genotype (*continued*)
of unrelated pairs, 136
Genotype correlation, 124
Genotype variable, 10–11, 13
as a construct, 11
IQ heritability definition and, 16
Gillie, O., 34, 38
Gini's formula, 236
Goddard, Henry, 8, 131
Goldberger, A. S., 6, 38, 118, 119, 148, 150,
202, 203, 204, 241
Honolulu group and, 175, 176, 180, *183,*
185, 186, 187, 188, 192, 197, 200,
201, 210, 213
Jensen's kinship equation and, 173, 174
Goldfeld, S. M., 192
Goodenough, D. R., 219, 222
Goodness of fit (Honolulu group), 186–
187, 192, 193–194, 196
Gordon, R. A., 131
Gottesman, I. I., 159, 216
GQ OPT program (Princeton University),
192
Graduate Record Examination (GRE), 229
Graybill, F. A., 162
Gregor, A. J., 67
"Group test" (Burt), 23, 35
Guilford, J. P., 131
Guthrie, L. F., 131

H

Haggard, E. A., 17
Half-siblings, 26. *See also* Siblings
Hall, W. S., 131
Haller, N. H., 8
Harrison, D. S., 131
Hart, H., *47,* 59
Heise, D. R., 119, 133
Hemispheric differentiation of the brain,
223–224. *See also* Brain
Henderson, N. D., 157
Heritability index (Falconer), 42, 148, 167
Heritability of IQ. *See* Adult IQ heritabil-
ity; Child IQ heritability; IQ her-
itability
Heritability ratio (Nichols), 42, 149, 167
Heritance index (Holzinger), 149, 167
Herman, L., *46,* 49, 54, *57*

Herrnstein, Richard, 6, 16, 72, 87, 110–112,
137, 157, 203, 204, 206, 211, 213,
223, 230
Burt scandal and, 39
heritability estimation and, 115, 116,
118, 120, 121, 126
Jencks's path estimates and, 103
kinship data and, 29, 32
linguists-IQ tests and, 131
main points of, 4–5
separated identical (MZ) twin correlation and, 18–19, 21, 76
Herrnstein's syllogism, 4
Heston, L. L., 159, 216
Hildreth, G., 32, *47*
Hirsch, J., 7, 156
Hogarth, R. M., 115, 123, 139, 143, 162, 174,
210
Hogben, L., *46,* 49, 54, *57*
Holzinger, K. J., 13, *46,* 82, 231
heritance index of, 42, 149, 167
Honolulu group, 6, 116, 119, 125, 126, 139,
205
adult and child IQ environmentability
and heritability and, 175–176,
187–192, 204, 205, 208
data sets of, 200–201
equations of, and modification, *182–*
185, 187–192
illustrative analysis of estimates of,
192–198
kinship data and, 27–29
models—procedures of, conclusions
concerning (Taylor), 198–201
overview of research of, 175–187
See also Morton, N.; Rao, D. C.
Horn, J. M., 149
Hornung, C. A., 162
Howard, Margaret, 15, 32, 36, 37, 38, 41,
121
Hudson, L., 35
Huntley, R. M. C., 50, 149
Husen, T., 50, 57, 149

I

"Individual test" (Burt), 23, 35, 52, 53
Inferential variable. *See* Constructs

Inflation of IQ correlation. *See* Artificial inflation of IQ correlation
Institute of Human Genetics (University of Copenhagen), 84
Intelligence
    chemical environment–genetic codes and, 27
    cognitive styles and, 219
    "general," 113, 131
    genotype, 11, 65–66, 109, 120, 159, 169, 233
    Honolulu group and, 177
    indexes of, 215
    IQ as different from, 12
    IQ tests and measuring of, 16
    mental age–chronological age and testing for, 102
    nature-nurture puzzle and, 1, 8, 199
    as unidimensional, 113
Interval scales, 130–131
Intraclass correlation coefficient, 17
Intraclass correlations, 152
Intrapair differences, general population and, 80
Intrapair environmental similarity. *See* Environmental similarity, intrapair
Intrauterine environment, 14, 18, 19, 125, 128–129, 134, 166–167, 211, 236
    Jencks's path estimates and, 104, 105, 106
Iowa Test of Basic Skills, 49, 51, 59
IQ (intelligence quotient)
    as different from intelligence, 12
    original conception of, 102
    reliability of, in time frame, 113
IQ game, defined, 7
IQ heritability
    bogus, kinship correlation and, 137–154, 207
    Burt scandal and, 34–41
    coefficient of heritability (h^2) and, 10–16
    collateral kinship model and, 124–137
    cross-validated (Jensen), 44, 56
    definition of, 16
    estimating, 16–18
        kinship correlation and, 25–34
        separated identical twin (MZA) correlation and, 18–25

IQ heritability, estimating (*continued*)
    estimation of, requirements for, 112–119
    estimation formulas of (in literature), 42
    fundamental equation and, 129, 134, 166
    gene-environment covariance, gene-environment interaction and, 68–72
    Honolulu research on
        conclusions concerning (Taylor), 198–200
        data sets of, 200–201
        equations and modification in, 187, 192
        illustrative analysis and, 192–198
        overview of, 175–187
    indeterminacy of, 170
    Jensen's kinship equation and, 41–64
    myth of separated identical (MZ) twins and, 75–111
    nature-nurture puzzle and, 1, 8, 199
    path analysis strategy and, 119–124
    problems remaining, 214–217
    race differences, between group heritability and, 64–68
    strategies for measuring, 16–18
    substantial evidence lacking for, 206–208
    as unreliable estimable quantity, 205–206
IQ heritability studies, 1–9, 112
    assumptions underlying, 209–212
    Birmingham group, 6, 29, 175, 186, 202–204, 205
    Burt scandal and, 34–41
    Honolulu group, 6, 27–29, 116, 119, 125, 126, 139, 175–201
    separated identical (MZ) twins and, 75–111
    shortcomings in, 212–214
IQ tests, cultural bias-dimensionality of, 131, 169. *See also* Test-retest reliability; Tests, unreliability of; *specific IQ tests*, e.g., Stanford-Binet IQ test; Wechsler Adult Intelligence Scales (WAIS), *etc.*
IQ variance equations, 122–123

Iterative weighted least squares (Hono-lulu group), 192

J

James, William, 220
Jarvik, L. F., 72, 118, 140, 241
   Jensen's kinship equation and, 43, 45, 53, 55, 171
   kinship data and, 29, 30, 31, 32, 33
Jencks, Christopher, 4, 15, 66, 70, 73, 137, 142, 143, 150, 157, 169, 213, 235, 236, 238, 240, 241
   causal sibling effect and, 127
   cross-pair correlation and, 126
   Honolulu group and, 178, 187, 190, 191, 200, 201
   Jensen's kinship equation and, 41, 42–43, 44, 45, 46, 48, 49, 50, 51, 52–55
   kinship data and, 34, 187
   path analysis of IQ inheritability and, 6, 119, 120, 123, 125
   separated twins and, 76
      Jencks's path estimates and, 103–109, 118
   specification errors and, 117–118
Jencks-Moore means, 55
Jensen, Arthur R., 9, 16, 34, 130, 137, 139, 148, 149, 150, 162, 168, 169, 203, 204, 206, 209, 210, 213, 235, 241
   additivity hypothesis of Jinks and Fulker and, 160
   analysis-of-variance strategy of, 119
   behavioral genetics and, 10, 13
   Burt scandal and, 36, 37, 38, 39, 41
   changes in IQ, environment and, 13–14
   collateral kinship and, 125, 126
   difference correlation of, 80
   dimensionality of IQ tests and, 131
   DZ twins and, 140
   errors of, 14, 230–231
   Eysenck and, 5
   families, IQ and, 12
   genetic makeup, intelligence and, 1–3, 4
   genetic strain (mice and rats) and, 157
   Gottesman and, 159

Jensen, Arthur R. (*continued*)
   heritability definition and, 13
   heritability estimation and, 116, 118, 120, 121, 123, 126
   Herrnstein's "meritocracy" and, 5
   interval scales and, 130
   kinship data of, 28, 29, 31, 32, 34
   kinship equation of, 41–64, 72, 73, 141, 167, 206
   "level" principle of, 132
   mean-to-difference correlation and, 161
   methodologies of, 18
   partitioning of environmental coeffi-cient, 15
   race differences and, 67
   residual variable and, 115
   separated twin correlation and, 18, 21, 22, 23, 24, 75, 76, 86–87, 88, 90, 93, 100, 103, 109, 110, 134, 135
      difference correlation of, 80, 81
      Juel-Nielsen data and, 83–84
      relatedness of adoptive families and, 91
      solving three kinship equations and, 171–175, 187, 207–208
   Stanford-Binet scores and, 36
Jensen's kinship equation, 41–64, 72, 73, 141, 167, 206
Jews, 8, 131
Jinks, J. L., 6, 50, 63, 71, 161, 202, 203, 239
   additivity hypothesis of, 159–160
   kinship data of, 29
Johnson, R. C., 86
Jones, H. E., 46, 60, 213
Jordan, W. D., 8
Jöreskog, K. G., 187
Juel-Nielsen, N., 212, 213, 237
   coded data of, 228
   separated MZ twins study of, 20, 21, 24, 33, 76, 77, 83–84, 85, 88, 89, 90, 91, 92, 100, 101, 102, 110, 206
   social environment data of, 93, 95, 99
Julesz, B., 224

K

Kagan, J., 131, 220, 221, 222
Kamin, L. J., 8, 21, 23, 54, 58, 67, 75, 118, 128, 140, 208, 212, 235, 236

Kamin, L. J. (*continued*)
  age-sex test standardization and, 80, 81,
      83, 101, 102
  birth weight and, 128
  Burt scandal and, 34, 35, 36, 37
  Honolulu group and, 187, 190, 200
  kinship data and, 31, 32
Karn, M. N., 33, 45, 49, 57, 59, 149
Kendall, M. G., 233, 236
Kerlinger, F. N., 113, 119, 162
Kimura, D., 223, 224
King, avoidance behavior and, 158
Kinometrics research, 6, 167, 213, 215
  defined, 149–150
Kinship correlation, 37, 170
  bogus heritability and, 137–154
  collateral kinship model and, 125
  estimating IQ heritability and, 18, 25–
      34
Kinship equation (Jensen's). *See* Jensen,
      Arthur R., kinship equation of
Kinships, direct line, 12, *28*, 177
Kinships, the study of
  many at a time
    Birmingham models and, 202–204
    equations of Honolulu group, and a
        modification, 187–192
    illustrative analysis of Honolulu esti-
        mates and, 192–198
    models and procedures of Honolulu
        group and, 198–201
    overview of Honolulu research on,
        175–187
    solving three equations, Jensen's ex-
        ample and, 171–175
  two at a time, 167
    bogus heritability, kinship correlation
        and, 137–154
    gene-environment interaction and,
        154–166
    general collateral kinship model and,
        124–137
    heritability estimation requirements
        and, 112–119
    pairwise comparisons of, 139, 148–
        154
    path analysis and, 119–124
Koch, H. L., 60, 129
Kogan, N., 219

L

Land, K. C., 119, 122
Late separation, MZ twin studies, 77,
      85–87
Law School Admission Test (LSAT), 229
Layzer, D., 15, 75, 130, 136, 137, 172
Leahy, A. M., 47, 52, 59, 107, 137, 191, 201
Least squares procedures
  Birmingham group and, 202
  Honolulu group and, 187
Lee, D., 221
Leik, R. K., 233
"Level" principle of Jensen, 132
Levy, J., 219, 223, 224
Lewontin, R. C., 6, 14, 65, 113, 156, 158
  Jensen's kinship equation and, 173
Light, R. J., 239
Lindzey, G., 41, 135
Linearity, 134, 137, 166, 212
  environmental differences and, 161
  heritability estimation and, 116–117
  Honolulu models and, 175, 180
Linear structural (multiple regression)
      equation, 121
Linguistic style differences, 131
Linn, R. L., 119
Loehlin, J. C., 17, 41, 50, 66, 67, 76, 105, 127,
      141, 161, 190, 191, 201, 202, 203,
      210, 213, 240
  specification errors and, 117
Lorge-Thorndike IQ test, 1
Luria, A. R., 219, 220
Lush, J. L., 13, 16
Lysenko, T. D., 110

M

McCall, R. B., 113
McClearn, G. E., 13, 22, 50, 64, 66, 86, 110,
      121, 149, 162
  kinship data and, 29
McClelland, D., 131, 219, 220, 221, 222
McNemar, Q., 21, 22, 47
Madsen, I. N., 47
Markus, G. B., 127
Marsh, F. J., 223, 224
Marwell, G., 162
Mazur, A., 243

Mean absolute intrapair differences, 237
Mean-to-difference correlation, 160
Measurement coefficient, 132
Measurement coefficient for environmental index, 177
Measurement errors, 81, 165
  collateral kinship model and, 130–134
  conclusions concerning (Taylor), 211–212
  IQ variance and, 123
  random, 132, 137, 169, 212
  residual variance and, 115
  in studies of MZ twins, 101–102
  systematic (nonrandom), 22, 115, 132
  of tests in question, 113
  variance due to, 14, 15
Mendelian "segregation," 180
Mental age, 102
Meritocracy, 5
Midparent, defined, 151–152
Miele, F., 67
Mill Hill Vocabulary Scale. See Raven's Mill Hill Vocabulary Scale
Milner, B., 224
Monozygotic-identical twins. See Twins, MZ (monozygotic-identical)
Morgensen, kinship data and (with Juel-Nielsen), 33
Morton, N., 6, 70, 123, 125, 136, 139, 148, 202, 203, 204, 208, 213, 217, 239, 240, 241
  Jencks's path estimates and, 103, 105
  Jensen's kinship equation and, 44, 50
  kinship data of, 29, 34, 175–201
  separated identical (MZ) twin correlation and, 76
Muller, H. J., 235
Multicollinearity, 14
Munsinger, H., 81, 128, 212
Myth of separation of identical twins. See Twins, MZ (monozygotic-identical), separation myth of
MZ twins. See Twins, MZ (monozygotic-identical)

N

Nachson, J., 224

Nader Report on Educational Testing Service, 229
Narrow heritability, 16, 203, 229. See also Broad heritability
National Merit Scholarship (vocabulary) Qualifying Test, 46, 50
Nature-nurture puzzle, 1, 8
  Honolulu group and, 199
Nebes, R. D., 224
Newell, environmental effects-genetic strains and (study-1967), 158
Newman, H. H., 13, 46, 53, 190, 193, 213, 216, 231, 236, 237, 238
  coded data of, 226
  separated twin study of, 17, 20, 21, 22, 76, 81–83, 84, 85, 86, 87, 88, 89, 90, 92, 100, 101, 102, 103, 106, 110, 206
  social environment data of, 93, 94–95, 97–99
Nichols, P. L., 45, 46, 50, 66, 127, 201, 235
  heritability index of, 42, 149, 167
Nonadditive path model illustration, 162–166. See also Path analysis of IQ heritability
Nonfamily environment. See Environment, specific (nonfamily)
Nongenetic differences. See Treatment effects (nongenetic differences) defined
Nonlinearity, 115, 123, 165. See also Linearity
Nonrecursive models, 216. See also Recursiveness
Nuclear families, 177, 179

O

Observed pair correlation, 132
Occupational categories (Burt's), 39–40
Olneck, M. R., 127
Ornstein, Robert, 218, 223
Osborn, R. T., 67
Osofsky, G., 8
Otis Advanced test, 46
Outhit, M. C., 47, 200
Overdetermination. See Overidentification (equation system)

Overidentification (equation system), 138
   Birmingham group and, 202
   Honolulu group and, 175, 187, 198

P

Pair-wise equation (Jensen-like), 215–216
Parameters (derived). *See* Derived parameters (Honolulu group)
Parameters (free). *See* Free parameters (Honolulu group)
Parent-offspring correlations, estimating IQ heritability and, 18
Partanen, J., 149
Partial regression coefficients, defined, 120
Path analysis of IQ heritability, 6, 118, 166, 231, 232
   collateral kinship model and, 126
   compound paths and, 122
   endogenous variables and, 121
   Honolulu group and, 177, 188
   Jencks and, 6, 103–109, 118, 119, 120, 123, 125
   kinship studies (two at a time) and, 119–124
   nonadditive path model illustration and, 162, 166
   path models and, 162, 177
   Wright's, 128
Path coefficient, 127, 234
   defined, 120, 121
   Honolulu group and, 186, 198
Path estimation equations, 122
Pearson, Karl, 7–8
Pearsonian correlation, 17
Pedhazur, E. J., 119, 162
Perceptual constancy, 67
Perturbations (Honolulu group), 199
Phenotype, 10–11, 16, 154
   birth weight, 104, 109, 128
   "intelligence," 215
   Jensen's kinship equation and, 42, 69
   measuring of, 12–13
Phenotype variables, 10–11, 13, 140
   as a construct, 11
   as multiplicative function of genotype and environment, 162

Phenotypic asymmetry (MZ twins). *See* Twins, MZ (monozygotic-identical), phenotypic asymmetry in
Phenotypic similarities, as reflection of IQ similarity, 16, 141
Phenylketonuria, 209
Phenylthiocarbamide (PTC), ability to taste, 12, 79, 81, 84
Piagetian perceptual constancy test, 67
Pintner, R., 33, 45, 49, 57
Plomin, R., 69, 71, 161, 162
Policy implications of IQ research. *See* Social policy implications of IQ research
Poniatoff, Alexander M., 3
Pooled correlations, 212
   pairs raised together, 45
Population Genetics Laboratory (University of Hawaii), 175. *See also* Honolulu group
Populations, 137, 205, 214
   linguistic style differences between, 131
   measured IQ for given, 112–113
   random samples and, 78
   sample bias and, 113, 211
   standard deviation for general, 80–81
   studies, heritability and, 13
   total, variances in IQ and, 14
   white, Jensen's kinship equation and, 44
Population variance, 115
Postnatal asymmetries, 167, 212
Predicted correlations (Honolulu group), 192
Prell, D. B., 46, 59
Prenatal environment. *See* Intrauterine environment
Project talent test, 46, 51

Q

Quandt, R. E., 192
Quetelet, Lambert Adolphe Jacques, 7, 39

R

Race, 44, 59, 74, 151, 169, 209, 229, 230
   cultural bias of IQ tests and, 131

Race (*continued*)
  difference, between-group heritability
    and, 64–68
  gene-environment covariance, gene-
    environment interaction and,
    71–72
  heritability of IQ and, 1–3, 5
    historical notes on, 7–9
  measurement bias and, 115
  measurement errors and, 113
Racial admixture, 67, 74
Ramirez, M., 223
Random mating, 202
Rao, D. C., 6, 70, 123, 125, 136, 202, 203,
    204, 208, 213, 239, 240, 241
  equations, 182–185
  Jencks's path estimates and, 103, 105
  Jensen's kinship equation and, 44, 50
  kinship data of, 27, 34, 118, 175–201
  separated identical (MZ) twin correla-
    tion and, 76
Raven's Mill Hill Vocabulary Scale, 20,
    127, 211
  composite scores and, 236
  Shields's study and, 23, 80, 160, 162
Raven's Progressive Matrices, 2, 20
  Shields's study and, 23
Reciprocal causation, 239
Recursiveness, 106, 120, 134, 137, 166, 169,
    211, 239
  definition of, 116–117
  Honolulu models and, 175, 180, 198
  See also Nonrecursive models
Relatedness of adoptive families, MZ twin
    studies and, 77, 90–92, 100–101,
    110, 237. *See also* Environmental
    similarity
Residual variables (nonfamily environ-
    ments), 105, 134, 138, 166
  heritability estimation and, 115–116,
    117, 130, 211
  Honolulu group and, 177, 181
Restriction range, 152, 153
  defined, 21
  Honolulu group and, 188
Reunion prior to testing, MZ twin studies
    and, 77, 87–90, 100, 110, 237. *See
    also* Environmental similarity
Robertson, L. S., 243

Roe, A., 236
Rubin, D. B., 39
Rudert, E. E., 131

S

Same-examiner effect, 101–102
Sample bias, 106, 110, 166, 169, 211
  cultural, 131
  heritability estimation and, 113, 115
  Juel-Nielsen and, 83–84
  Newman and, 81–83
  Shields and, 78–81
  systematic, 133
Samples, R. E., 223
Samuda, R. J., 131
Sanday, P. R., 58
Sasaki, M. S., 162
Saudek, R., 236
Scarr, S., 7, 60, 68, 107, 128, 143, 150–151,
    152–154, 167, 206, 208, 210, 212,
    216
  Honolulu group and, 177, 187, 191, 201
Scarr-Salapatek, S., 4, 7, 51, 56, 57, 59, 60,
    143, 150, 159, 212, 216
  gene-environment interaction and, 71
  heritability values for blacks and, 66–
    67
Schoenfelt, kinship studies used by Jencks
    and, 46
Scholastic Aptitude Test (SAT), 229
Schwartz, J., 63, 128, 148, 161
Schwartz, Michael, 63, 128, 148, 161, 234
Second cousins, 26, 28, 34
Segall, M. H., 243
Selective placement, 152, 153, 210
  Honolulu group and, 188, 190–191
Separated identical (MZA) twin correla-
    tion, 18–25
  Burt scandal and, 35, 36, 37, 38
  See also Twins, MZ (monozygotic-iden-
    tical), separation myth of
Sex differences, cognitive style and, 221
Sex standardization of tests, 80, 81, 83, 101,
    102, 113
Shields, J., 212, 216, 235, 236, 237
  Burt scandal and, 38
  coded data of, 227

Shields, J. (*continued*)
  separated MZ twin study of, *20*, 21, 22–23, 76, 77, 78–81, 84, 85, 86, 87, 88, *89*, 90, 91, 92, 98, 100, 101–102, 110, 129, 160, 162
  social environment data of, 93, 94, 95–97
  transformed scores of, 236
Shockley, William, 3, 6, 130, 209
Shuey, A., 5, 9
Siblings, 11, 17, 74, 137, 140, 204, 207, 214
  Burt scandal and, 37
  gene-environment correlation and, 69, 127
  IQ test scores and, 12
  kinship correlation and, 26, 27, *28*, 32, 33
  natural, 152, 153
  raised together, Jensen's kinship equation and, 42–44, 45, *46, 48, 49*, 50, 52, 53–54, *56, 57*, 58, 61–62, 72
Similarity in social environment of MZ twins. *See* Environmental similarity of separated identical (MZ) twins
Skeels, H. M., 107, 191, 201
Skodak, M., *47*, 59, 107, 137, 191, 200, 201, 208
Smith, K. W., 162
Smith, P. V., 239
Smith, R. T., 60
Snider, B., 33, 45, 49, *57*
Snygg, D., 208
Social environment, 1, 27, 77, 134, 151, 202, 237
  Herrnstein and, 4
  Jencks's path estimates and, 105
  Jensen's kinship equation and, 60
  kinship correlation and, 27
  similarity in, of MZ twins, 92–110
  See also Environmental similarity
Social learning hypothesis, 220–221
Social policy implications of IQ research
  cognitive style and, 225
  genetic research and, 112, 130
  IQ inheritability theory and, 3
Socioeconomic status (SES), 177, 198, 201, 214, 215
Southwood, K. E., 162

Specific (noncommon) environment. *See* Environment, specific (noncommon)
Specification errors, 117–118
Sperry, Roger, 218, 223
Spielman, W., 39
Spinnler, H., 224
Split-brain theory, cognitive style and, 223–224. *See also* Brain
Spuhler, J. N., 41, 135
Standard deviation, 236, 237. *See also* Populations, standard deviation for general
Stanford-Binet IQ test, 1, *46*, 49, 51, 54, 59, 113, 221
  bias in, 131
  Burt's study and, 23–24, 36, 39
  Herrnstein and, 4
  Jews and, 8
  McNemar's incorrect correlation and, 22
  Newman's study and, 89, 97, 102
  race differences and, 66
  separated MZ twin studies and, *20*
  Shields's study and, 23
Sterilization, 3
Stigler, S. M., 39
Stinchcombe, A., 6
Stocks, P., 33, 45, 49, *57*, 59, 149
Structural equations (behavioral sciences), 121, 162
Structure-of-intellect model, dimensionality of, 131

T

Tanser, H. A., 67
Taubman, P., 6, 149–150, 213
Taylor, H. F., 120, 123, 162
TenHouten, W. D., 223
Terman, L. F., 52, 200, 213
Test-retest correlation, 132
Test-retest reliability, 53, 83, 130
Tests, unreliability of, 14
  corrections for, 54, 55
  See also Measurement error
Thurstone PMA test, *46*, 51, 59
Tirre, W. C., 131

Trabasso, T., 131
Treatment effects
  environmental correlation and, 148
  nongenetic differences, defined, 63
  See also Environmental treatments
Twins
  DZ (dizygotic-fraternal), 31, 74, 79, 137,
      139, 140, 141, 148, 149–150, 167,
      168, 204, 210, 214–215, 231, 232–
      233
    Burt scandal and, 37
    comparing with MZ, 26–27
    defined, 25–26
    gene-environment correlation and,
        69
    Honolulu group and, 193–194
    Jensen's kinship equation and, 41–64
    kinship data and, 28, 31, 32–33
    race differences and, 66
  gene-environment correlation and, 127
  MZ (monozygotic-identical), 74, 137,
      141, 143, 148, 149–150, 160, 168,
      169, 170, 204, 210, 214–215, 231
    Burt scandal and study of, 35, 36, 37,
        38
    collateral kinship model and, 134–135
    defined, 18
    estimating IQ heritability and, 206–
        207, 213, 216
      collateral kinships and, 11, 72
      IQ correlation for, 18
      IQ scores and, 12
      kinship correlation and, 25–26, 27,
          28, 33
      separated identical twin (MZA)
          correlation and, 18–25
    gene-environment correlation and,
        69
    Jensen's kinship equation and, 41–64
    separation myth of
      analysis of data and, 84–102
      environmental similarity data and,
          92–101
      Honolulu group and, 190, 191, 193–
          194, 199
      Jencks's path estimates and, 103–
          109
      late separation data and, 85–87
      measurement error and, 101–102

Twins (*continued*)
    overview of, 75–78, 109–111
    relatedness of adoptive families
        data and, 90–92
    reunion prior to testing data and,
        87–90
    sample-sample bias and, 78–84
  study of, by Galton, 8

U

Uncorrected for reliability (Burt), 55
Uncorrelated residuals, 134, 137
Underidentification (equation system),
    138. *See also* Overidentification
    (equation system)
Unrelated individuals, 26, 137
  adopted-raised together, Jensen's kin-
      ship equation and, 42, 43, 45, 46,
      48, 49, 52, 55, 56, 57, 58, 62–63,
      135–137, 170
  children, raised together, 46, 48, 54, 56
  kinship studies and, 31, 33
  pairs, gene-environment correlation
      and, 127
Uterine environment. *See* Intrauterine en-
    vironment

V

Van Allen, M. W., 224
Vandenberg, S. G., 50, 57, 63, 67, 86, 149,
    213
Vasgird, Daniel R., on cognitive style, 218–
    225
Vocabulary tests, 131. *See also* Raven's
    Mill Hill Vocabulary Scale
Vogel, F., 149

W

Wade, N., 34, 35
Watson, John B., 83, 236
Wechsler Adult Intelligence Scales
    (WAIS), 1, 113, 211
  Herrnstein and, 4
  Juel-Nielsen study and, 24, 83, 101, 102
  Shields's study and, 23

Wechsler-Bellevue IQ test, 20, 46
Weinberg, R. A., 7, 107, 143, 150–151,
       152–154, 167, 206, 208, 210, 212,
       216
   Honolulu group and, 177, 187, 191, 201
Werts, C. E., 119
Whites. See Race
Wictorin, M., 50, 57, 63, 149
Willerman, L., 102
Williams, R. L., 131
Willoughby, R. R., 54, 56, 200, 213
Wilson, P. T., 60
Within-class variance, 17
Within-group differences in IQ heritabil-
       ity, 9, 209
Within-group heritability, 64
Witkin, Herman, 131, 218, 219, 220–221,
       222, 223, 225

Wright, Sewall, 70, 119, 120, 122, 128, 166,
       175, 202

Y

Yee, S., 199
Yin, R. K., 224
Yule, G. U., 233

Z

Zajonc, R. B., 127
Zazzo, R., 149
Zero-order (or total) coefficient, 121
Z-transforms, 89, 91, 120, 242
Zygosity of MZ twin pairs, 79, 81–82, 84